These safety symbols are used in laboratory and field investigations in thi[s]
ing of each symbol and refer to this page often. *Remember to wash your [...]*

## PROTECTIVE EQUIPMENT  Do not begin any lab without the pr[...]

 **GOGGLES** Proper eye protection must be worn when performing or observing science activities which involve items or conditions as listed below.

 **APRON** Wear an approved apron when using substances that could stain, wet, or destroy cloth.

 **SOAP** Wash hands with soap and water before removing goggles and after all lab activities.

 **GLOVES** Wear gloves when working with biological materials, chemicals, animals, or materials that can stain or irritate hands.

## LABORATORY HAZARDS

| Symbols | Potential Hazards | Precaution | Response |
|---|---|---|---|
| **DISPOSAL** | contamination of classroom or environment due to improper disposal of materials such as chemicals and live specimens | • DO NOT dispose of hazardous materials in the sink or trash can.<br>• Dispose of wastes as directed by your teacher. | • If hazardous materials are disposed of improperly, notify your teacher immediately. |
| **EXTREME TEMPERATURE** | skin burns due to extremely hot or cold materials such as hot glass, liquids, or metals; liquid nitrogen; dry ice | • Use proper protective equipment, such as hot mitts and/or tongs, when handling objects with extreme temperatures. | • If injury occurs, notify your teacher immediately. |
| **SHARP OBJECTS** | punctures or cuts from sharp objects such as razor blades, pins, scalpels, and broken glass | • Handle glassware carefully to avoid breakage.<br>• Walk with sharp objects pointed downward, away from you and others. | • If broken glass or injury occurs, notify your teacher immediately. |
| **ELECTRICAL** | electric shock or skin burn due to improper grounding, short circuits, liquid spills, or exposed wires | • Check condition of wires and apparatus for fraying or uninsulated wires, and broken or cracked equipment.<br>• Use only GFCI-protected outlets | • DO NOT attempt to fix electrical problems. Notify your teacher immediately. |
| **CHEMICAL** | skin irritation or burns, breathing difficulty, and/or poisoning due to touching, swallowing, or inhalation of chemicals such as acids, bases, bleach, metal compounds, iodine, poinsettias, pollen, ammonia, acetone, nail polish remover, heated chemicals, mothballs, and any other chemicals labeled or known to be dangerous | • Wear proper protective equipment such as goggles, apron, and gloves when using chemicals.<br>• Ensure proper room ventilation or use a fume hood when using materials that produce fumes.<br>• NEVER smell fumes directly.<br>• NEVER taste or eat any material in the laboratory. | • If contact occurs, immediately flush affected area with water and notify your teacher.<br>• If a spill occurs, leave the area immediately and notify your teacher. |
| **FLAMMABLE** | unexpected fire due to liquids or gases that ignite easily such as rubbing alcohol | • Avoid open flames, sparks, or heat when flammable liquids are present. | • If a fire occurs, leave the area immediately and notify your teacher. |
| **OPEN FLAME** | burns or fire due to open flame from matches, Bunsen burners, or burning materials | • Tie back loose hair and clothing.<br>• Keep flame away from all materials.<br>• Follow teacher instructions when lighting and extinguishing flames.<br>• Use proper protection, such as hot mitts or tongs, when handling hot objects. | • If a fire occurs, leave the area immediately and notify your teacher. |
| **ANIMAL SAFETY** | injury to or from laboratory animals | • Wear proper protective equipment such as gloves, apron, and goggles when working with animals.<br>• Wash hands after handling animals. | • If injury occurs, notify your teacher immediately. |
| **BIOLOGICAL** | infection or adverse reaction due to contact with organisms such as bacteria, fungi, and biological materials such as blood, animal or plant materials | • Wear proper protective equipment such as gloves, goggles, and apron when working with biological materials.<br>• Avoid skin contact with an organism or any part of the organism.<br>• Wash hands after handling organisms. | • If contact occurs, wash the affected area and notify your teacher immediately. |
| **FUME** | breathing difficulties from inhalation of fumes from substances such as ammonia, acetone, nail polish remover, heated chemicals, and mothballs | • Wear goggles, apron, and gloves.<br>• Ensure proper room ventilation or use a fume hood when using substances that produce fumes.<br>• NEVER smell fumes directly. | • If a spill occurs, leave area and notify your teacher immediately. |
| **IRRITANT** | irritation of skin, mucous membranes, or respiratory tract due to materials such as acids, bases, bleach, pollen, mothballs, steel wool, and potassium permanganate | • Wear goggles, apron, and gloves.<br>• Wear a dust mask to protect against fine particles. | • If skin contact occurs, immediately flush the affected area with water and notify your teacher. |
| **RADIOACTIVE** | excessive exposure from alpha, beta, and gamma particles | • Remove gloves and wash hands with soap and water before removing remainder of protective equipment. | • If cracks or holes are found in the container, notify your teacher immediately. |

# Your online portal to everything you need

connectED.mcgraw-hill.com

Look for these icons to access exciting digital resources

 Video

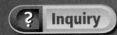 Audio

 Review

 Inquiry

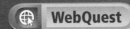 WebQuest

 Assessment

Concepts in Motion

MOTION AND FORCES

iSCIENCE

Glencoe

**MOTION AND FORCES**

**i SCIENCE**

Glencoe

**Bubbles**

The iridescent colors of these soap bubbles result from a property called interference. Light waves reflect off both outside and inside surfaces of bubbles. When this happens, the waves interfere with each other and you see different colors. The thickness of the soap film that forms a bubble also affects interference.

The **McGraw·Hill** Companies

 **Education**

Send all inquiries to:
McGraw-Hill Education
8787 Orion Place
Columbus, OH 43240-4027

ISBN: 978-0-07-888019-3
MHID: 0-07-888019-X

Printed in the United States of America.

5 6 7 8 9 10 11 12 13  DOW  15 14 13

## Authors

**American Museum of Natural History**
New York, NY

**Michelle Anderson, MS**
Lecturer
The Ohio State University
Columbus, OH

**Juli Berwald, PhD**
Science Writer
Austin, TX

**John F. Bolzan, PhD**
Science Writer
Columbus, OH

**Rachel Clark, MS**
Science Writer
Moscow, ID

**Patricia Craig, MS**
Science Writer
Bozeman, MT

**Randall Frost, PhD**
Science Writer
Pleasanton, CA

**Lisa S. Gardiner, PhD**
Science Writer
Denver, CO

**Jennifer Gonya, PhD**
The Ohio State University
Columbus, OH

**Mary Ann Grobbel, MD**
Science Writer
Grand Rapids, MI

**Whitney Crispen Hagins, MA, MAT**
Biology Teacher
Lexington High School
Lexington, MA

**Carole Holmberg, BS**
Planetarium Director
Calusa Nature Center and
Planetarium, Inc.
Fort Myers, FL

**Tina C. Hopper**
Science Writer
Rockwall, TX

**Jonathan D. W. Kahl, PhD**
Professor of Atmospheric Science
University of Wisconsin-
Milwaukee
Milwaukee, WI

**Nanette Kalis**
Science Writer
Athens, OH

**S. Page Keeley, MEd**
Maine Mathematics and
Science Alliance
Augusta, ME

**Cindy Klevickis, PhD**
Professor of Integrated Science
and Technology
James Madison University
Harrisonburg, VA

**Kimberly Fekany Lee, PhD**
Science Writer
La Grange, IL

**Michael Manga, PhD**
Professor
University of California, Berkeley
Berkeley, CA

**Devi Ried Mathieu**
Science Writer
Sebastopol, CA

**Elizabeth A. Nagy-Shadman, PhD**
Geology Professor
Pasadena City College
Pasadena, CA

**William D. Rogers, DA**
Professor of Biology
Ball State University
Muncie, IN

**Donna L. Ross, PhD**
Associate Professor
San Diego State University
San Diego, CA

**Marion B. Sewer, PhD**
Assistant Professor
School of Biology
Georgia Institute of Technology
Atlanta, GA

**Julia Meyer Sheets, PhD**
Lecturer
School of Earth Sciences
The Ohio State University
Columbus, OH

**Michael J. Singer, PhD**
Professor of Soil Science
Department of Land, Air and
Water Resources
University of California
Davis, CA

**Karen S. Sottosanti, MA**
Science Writer
Pickerington, Ohio

**Paul K. Strode, PhD**
I.B. Biology Teacher
Fairview High School
Boulder, CO

**Jan M. Vermilye, PhD**
Research Geologist
Seismo-Tectonic Reservoir
Monitoring (STRM)
Boulder, CO

**Judith A. Yero, MA**
Director
Teacher's Mind Resources
Hamilton, MT

**Dinah Zike, MEd**
Author, Consultant,
Inventor of Foldables
Dinah Zike Academy;
Dinah-Might Adventures, LP
San Antonio, TX

**Margaret Zorn, MS**
Science Writer
Yorktown, VA

## Consulting Authors

**Alton L. Biggs**
Biggs Educational Consulting
Commerce, TX

**Ralph M. Feather, Jr., PhD**
Assistant Professor
Department of Educational
Studies and Secondary
Education
Bloomsburg University
Bloomsburg, PA

**Douglas Fisher, PhD**
Professor of Teacher Education
San Diego State University
San Diego, CA

**Edward P. Ortleb**
Science/Safety Consultant
St. Louis, MO

## Series Consultants

### Science

**Solomon Bililign, PhD**
Professor
Department of Physics
North Carolina Agricultural
and Technical State University
Greensboro, NC

**John Choinski**
Professor
Department of Biology
University of Central Arkansas
Conway, AR

**Anastasia Chopelas, PhD**
Research Professor
Department of Earth and
Space Sciences
UCLA
Los Angeles, CA

**David T. Crowther, PhD**
Professor of Science Education
University of Nevada, Reno
Reno, NV

**A. John Gatz**
Professor of Zoology
Ohio Wesleyan University
Delaware, OH

**Sarah Gille, PhD**
Professor
University of California
San Diego
La Jolla, CA

**David G. Haase, PhD**
Professor of Physics
North Carolina State
University
Raleigh, NC

**Janet S. Herman, PhD**
Professor
Department of Environmental
Sciences
University of Virginia
Charlottesville, VA

**David T. Ho, PhD**
Associate Professor
Department of Oceanography
University of Hawaii
Honolulu, HI

**Ruth Howes, PhD**
Professor of Physics
Marquette University
Milwaukee, WI

**Jose Miguel Hurtado, Jr.,
PhD**
Associate Professor
Department of Geological
Sciences
University of Texas at El Paso
El Paso, TX

**Monika Kress, PhD**
Assistant Professor
San Jose State University
San Jose, CA

**Mark E. Lee, PhD**
Associate Chair & Assistant
Professor
Department of Biology
Spelman College
Atlanta, GA

**Linda Lundgren**
Science writer
Lakewood, CO

## Series Reviewers, continued

**Carolyn Elliott**
Iredell-Statesville Schools
Statesville, NC

**Christine M. Jacobs**
Ranger Middle School
Murphy, NC

**Jason O. L. Johnson**
Thurmont Middle School
Thurmont, MD

**Felecia Joiner**
Stony Point Ninth Grade Center
Round Rock, TX

**Joseph L. Kowalski, MS**
Lamar Academy
McAllen, TX

**Brian McClain**
Amos P. Godby High School
Tallahassee, FL

**Von W. Mosser**
Thurmont Middle School
Thurmont, MD

**Ashlea Peterson**
Heritage Intermediate Grade Center
Coweta, OK

**Nicole Lenihan Rhoades**
Walkersville Middle School
Walkersvillle, MD

**Maria A. Rozenberg**
Indian Ridge Middle School
Davie, FL

**Barb Seymour**
Westridge Middle School
Overland Park, KS

**Ginger Shirley**
Our Lady of Providence
Junior-Senior High School
Clarksville, IN

**Curtis Smith**
Elmwood Middle School
Rogers, AR

**Sheila Smith**
Jackson Public School
Jackson, MS

**Sabra Soileau**
Moss Bluff Middle School
Lake Charles, LA

**Tony Spoores**
Switzerland County Middle School
Vevay, IN

**Nancy A. Stearns**
Switzerland County Middle School
Vevay, IN

**Kari Vogel**
Princeton Middle School
Princeton, MN

**Alison Welch**
Wm. D. Slider Middle School
El Paso, TX

**Linda Workman**
Parkway Northeast Middle School
Creve Coeur, MO

## Teacher Advisory Board

The Teacher Advisory Board gave the authors, editorial staff, and design team feedback on the content and design of the Student Edition. They provided valuable input in the development of *Glencoe ⓘScience*.

**Frances J. Baldridge**
Department Chair
Ferguson Middle School
Beavercreek, OH

**Jane E. M. Buckingham**
Teacher
Crispus Attucks Medical
Magnet High School
Indianapolis, IN

**Elizabeth Falls**
Teacher
Blalack Middle School
Carrollton, TX

**Nelson Farrier**
Teacher
Hamlin Middle School
Springfield, OR

**Michelle R. Foster**
Department Chair
Wayland Union
Middle School
Wayland, MI

**Rebecca Goodell**
Teacher
Reedy Creek Middle School
Cary, NC

**Mary Gromko**
Science Supervisor K–12
Colorado Springs District 11
Colorado Springs, CO

**Randy Mousley**
Department Chair
Dean Ray Stucky
Middle School
Wichita, KS

**David Rodriguez**
Teacher
Swift Creek Middle School
Tallahassee, FL

**Derek Shook**
Teacher
Floyd Middle Magnet School
Montgomery, AL

**Karen Stratton**
Science Coordinator
Lexington School District One
Lexington, SC

**Stephanie Wood**
Science Curriculum Specialist,
K–12
Granite School District
Salt Lake City, UT

# Online Guide

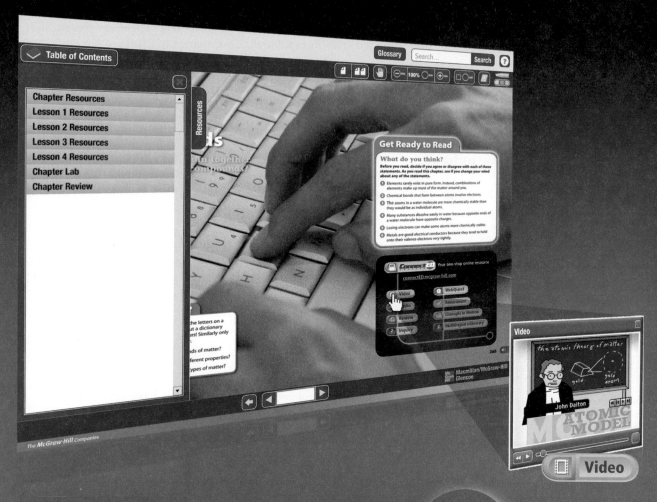

## Connect**ED**

▷ **Your Digital Science Portal**

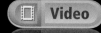 **Video**

See the science in real life through these exciting

 **Audio**

Click the link and you can listen to the text while you

 **Review**

Try these interactive tools to help you review

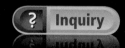 **Inquiry**

Explore concepts through hands-on and virtual labs

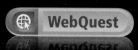 **WebQuest**

These web-based challenges relate the concepts you're learning

The icons in your online student edition link you to interactive learning opportunities. Browse your online student book to find more.

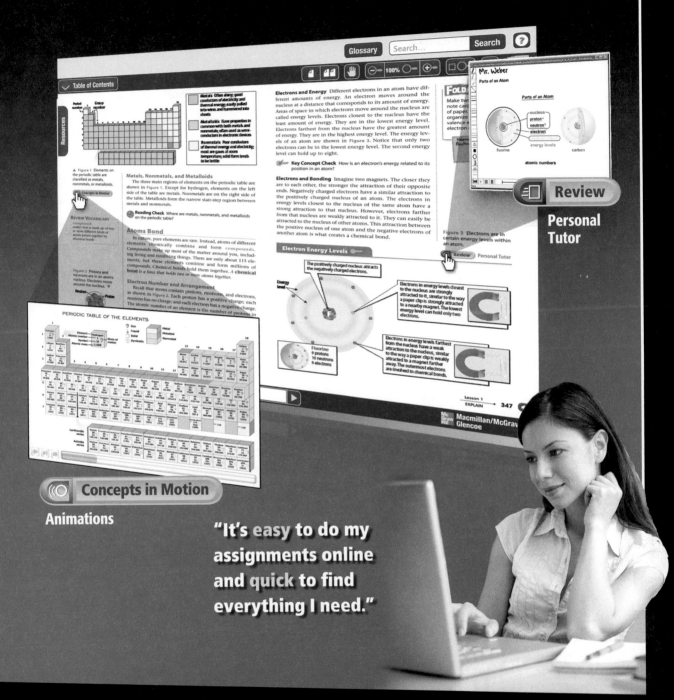

Concepts in Motion

**Animations**

"It's easy to do my assignments online and quick to find everything I need."

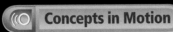

**Assessment**

Check how well you understand the concepts with online quizzes and practice questions.

**Concepts in Motion**

The textbook comes alive with animated explanations of important concepts.

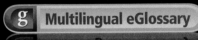

**Multilingual eGlossary**

Read key vocabulary in 13 languages.

# Treasure Hunt

Your science book has many features that will aid you in your learning. Some of these features are listed below. You can use the activity at the right to help you find these and other special features in the book.

- **THE BIG IDEA** can be found at the start of each chapter.

- The Reading Guide at the start of each lesson lists 🔑 **Key Concepts**, vocabulary terms, and online supplements to the content.

- **Connect ED** icons direct you to online resources such as animations, personal tutors, math practices, and quizzes.

- **Inquiry** Labs and Skill Practices are in each chapter.

- Your **FOLDABLES** help organize your notes.

**1** What four margin items can help you build your vocabulary?

**2** On what page does the glossary begin? What glossary is online?

**3** In which Student Resource at the back of your book can you find a listing of Laboratory Safety Symbols?

**4** Suppose you want to find a list of all the Launch Labs, MiniLabs, Skill Practices, and Labs, where do you look?

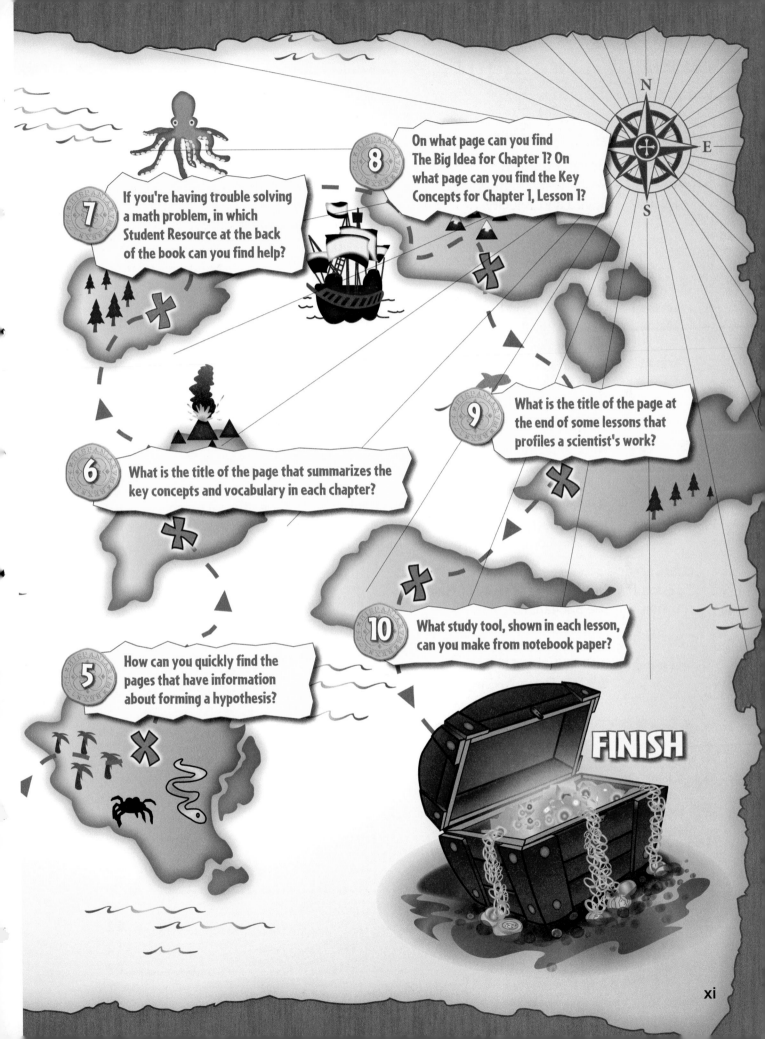

If you're having trouble solving a math problem, in which Student Resource at the back of the book can you find help?

On what page can you find The Big Idea for Chapter 1? On what page can you find the Key Concepts for Chapter 1, Lesson 1?

What is the title of the page at the end of some lessons that profiles a scientist's work?

What is the title of the page that summarizes the key concepts and vocabulary in each chapter?

What study tool, shown in each lesson, can you make from notebook paper?

How can you quickly find the pages that have information about forming a hypothesis?

FINISH

# Table of Contents

# Inquiry Labs

## Inquiry Launch Labs

## Inquiry MiniLabs

# Inquiry Labs

## Inquiry Skill Practice

## Inquiry Inquiry Labs

# Features

TABLE OF CONTENTS

**XV**

# Unit 1

## Motion & Forces

**3500 B.C.**
The oldest wheeled vehicle is depicted in Mesopotamia, near the Black Sea.

**400 B.C.**
The Greeks invent the stone-hurling catapult.

**1698**
English military engineer Thomas Savery invents the first crude steam engine while trying to solve the problem of pumping water out of coal mines.

**1760–1850**
The Industrial Revolution results in massive advances in technology and social structure in England.

**1769**
The first vehicle to move under its own power is designed by Nicholas Joseph Cugnot and constructed by M. Breszin. A second replica is built that weighs 3,629 kg and has a top speed of 3.2 km per hour.

**1794**
Eli Whitney receives a patent for the mechanical cotton gin.

**1817**
Baron von Drais invents a machine to help him quickly wander the grounds of his estate. The machine is made of two wheels on a frame with a seat and a pair of pedals. This machine is the beginning design of the modern bicycle.

**1903**
Wilbur and Orville Wright build their airplane, called the Flyer, and take the first successful, powered, piloted flight.

**1976**
The first computer for home use is invented by college dropouts Steve Wozniak and Steve Jobs, who go on to found Apple Computer, Inc.

**? Inquiry**
Visit ConnectED for this unit's **STEM** activity.

# Models

Have you ridden on an amusement park roller coaster such as the one in **Figure 1?** As you were going down the steepest hill or hanging upside down in a loop, did you think to yourself, "I hope I don't fly off this thing"? Before construction begins on a roller coaster, engineers build different models of the thrill ride to ensure proper construction and safety. A **model** is a representation of an object, an idea, or a system that is similar to the physical object or idea being studied.

**Figure 1** Engineers use various models to design roller coasters.

## Using Models in Physical Science

Models are used to study things that are too big or too small, happen too quickly or too slowly, or are too dangerous or too expensive to study directly. Different types of models serve different purposes. Roller-coaster engineers might build a physical model of their idea for a new, daring coaster. Using mathematical and computer models, the engineers can calculate the measurements of hills, angles, and loops to ensure a safe ride. Finally, the engineers might create another model called a blueprint, or drawing, that details the construction of the ride. Studying the various models allows engineers to predict how the actual roller coaster will behave when it travels through a loop or down a giant hill.

## Types of Models

### Physical Model

A physical model is a model that you can see and touch. It shows how parts relate to one another, how something is built, or how complex objects work. Physical models often are built to scale. A limitation of a physical model is that it might not reflect the physical behavior of the full-size object. For example, this model will not accurately show how wind will affect the ride.

### Mathematical Model

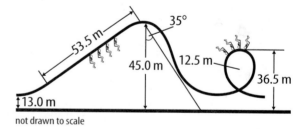

53.5 m  35°  45.0 m  12.5 m  36.5 m  13.0 m

not drawn to scale

A mathematical model uses numerical data and equations to model an event or idea. Mathematical models often include input data, constants, and output data. When designing a thrill ride, engineers use mathematical models to calculate the heights, the angles of loops and turns, and the forces that affect the ride. One limitation of a mathematical model is that you cannot use it to model how different parts are assembled.

## Making Models

An important factor in making a model is determining its purpose. You might need a model that physically represents an object. Or, you might need a model that includes only important elements of an object or a process. When you build a model, first determine the function of the model. What variables need to change? What materials should you use? What do you need to communicate to others? **Figure 2** shows two models of a glucose molecule, each with a different purpose.

## Limitations of Models

It is impossible to include all the details about an object or an idea into one model. All models have limitations. When using models to design a structure, an engineer must be aware of the information each model does and does not provide. For example, a blueprint of a roller coaster does not show the maximum weight that a car can support. However, a mathematical model would include this information. Scientists and engineers consider the purpose and the limitations of the model they use to ensure they draw accurate conclusions from models.

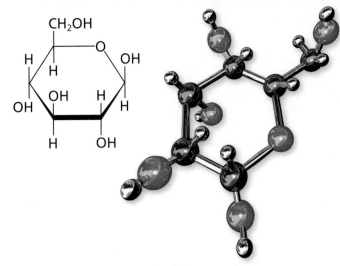

**Figure 2** The model on the left is used to represent how the atoms in a glucose molecule bond together. The model on the right is a 3-D representation of the molecule, which shows how atoms might interact.

## Computer Simulation

A computer simulation is a model that combines large amounts of data and mathematical models with computer graphic and animation programs. Simulations can contain thousands of complex mathematical models. When roller coaster engineers change variables in mathematical models, they use computer simulation to view the effects of the change.

---

### Inquiry MiniLab
**30 minutes**

#### Can you model a roller coaster?

You are an engineer with an awesome idea for a new roller coaster—the car on your roller coaster makes a jump and then lands back on the track. You model your idea to show it to managers at a theme park in hopes that you can build it.

1. Read and complete a lab safety form.
2. Create a blueprint of your roller coaster. Include a scale and measurements.
3. Follow your blueprint to build a scaled physical model of your roller coaster. Use **foam hose insulation, tape,** and other **craft supplies.**
4. Use a **marble** as a model for a roller-coaster car. Test your model. Record your observations in your Science Journal.

#### Analyze and Conclude

1. **Compare** your blueprint and physical model.

2. **Evaluate** After you test your physical model, list the design changes you would make to your blueprint.

3. **Identify** What are the limitations of each of your models?

# Describing Motion

 **THE BIG IDEA** **What are some ways to describe motion?**

## How is their motion changing?

Have you ever seen a group of planes zoom through the sky at an air show? When one plane speeds up, all the planes speed up. When one plane turns, all the other planes turn in the same direction.

- What might happen if all the planes did not move in the same way?
- How could you describe the positions of the planes in the photo?
- What are some ways you could describe the motion of the planes?

# Get Ready to Read

## What do you think?

Before you read, decide if you agree or disagree with each of these statements. As you read this chapter, see if you change your mind about any of the statements.

**1** Displacement is the distance an object moves along a path.

**2** The description of an object's position depends on the reference point.

**3** Constant speed is the same thing as average speed.

**4** Velocity is another name for speed.

**5** You can calculate average acceleration by dividing the change in velocity by the change in distance.

**6** An object accelerates when either its speed or its direction changes.

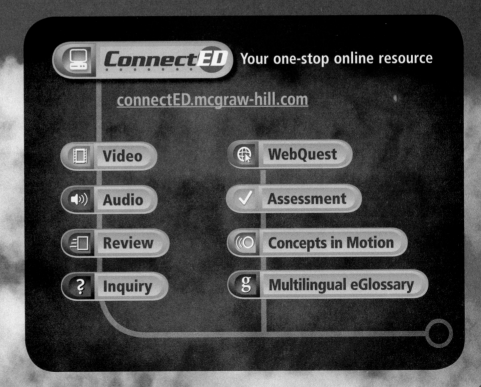

**ConnectED** Your one-stop online resource

connectED.mcgraw-hill.com

- Video
- WebQuest
- Audio
- Assessment
- Review
- Concepts in Motion
- Inquiry
- Multilingual eGlossary

# Lesson 1

## Reading Guide

### Key Concepts 🔑
**ESSENTIAL QUESTIONS**

- How does the description of an object's position depend on a reference point?
- How can you describe the position of an object in two dimensions?
- What is the difference between distance and displacement?

### Vocabulary
**reference point** p. 9
**position** p. 9
**motion** p. 13
**displacement** p. 13

 **Multilingual eGlossary**

# Position and Motion

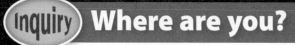 **Where are you?**

A short time ago, people on this ship probably saw only open ocean. They knew where they were only by looking at the instruments on the ship. But the situation has changed. How can the lighthouse help the ship's crew guide the ship safely to shore?

### How do you get there from here?

How would you give instructions to a friend who was trying to walk from one place to another in your classroom?

1. Read and complete a lab safety form.
2. Place a sheet of **paper** labeled *North, East, South,* and *West* on the floor.
3. Walk from the paper to one of the three locations your teacher has labeled in the classroom. Have a partner record the number of steps and the directions of movement in his or her Science Journal.
4. Using these measurements, write instructions other students could follow to move from the paper to the location.
5. Repeat steps 3 and 4 for the other locations.

**Think About This**

1. How did your instructions to each location compare to those written by other groups?

2. 🔑 **Key Concept** How did the description of your movement depend on the point at which you started?

## Describing Position

How would you describe where you are right now? You might say you are sitting one meter to the left of your friend. Perhaps you would explain that you are at home, which is two houses north of your school. You might instead say that your house is ten blocks east of the center of town, or even 150 million kilometers from the Sun.

What do all these descriptions have in common? Each description states your location **relative** to a certain point. *A* **reference point** *is the starting point you choose to describe the location, or position, of an object.* The reference points in the first paragraph are your friend, your school, the center of town, and the Sun.

Each description of your location also includes your distance and direction from the reference point. Describing your location in this way defines your position. *A* **position** *is an object's distance and direction from a reference point.* A complete description of your position includes a distance, a direction, and a reference point.

✔️ **Reading Check** What are two ways you could describe your position right now?

**SCIENCE USE V. COMMON USE**

**relative**
*Science Use* compared (to)

*Common Use* a member of your family

**Figure 1** The arrows indicate the distances and directions from different reference points.

✔️ **Visual Check** How do you know which reference point is farther from the table?

10 m

Entrance

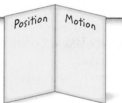

**FOLDABLES**

Fold a sheet of paper to make a half book. Use it to organize your notes about how position and motion are related.

Position | Motion

## Using a Reference Point to Describe Position

Why do you need a reference point to describe position? Suppose you are planning a family picnic. You want your cousin to arrive at the park early to save your favorite picnic table. The park is shown in **Figure 1**. How would you describe the position of your favorite table to your cousin? First, choose a reference point that a person can easily find. In this park, the statue is a good choice. Next, describe the direction that the table is from the reference point—toward the slide. Finally, say how far the table is from the statue—about 10 m. You would tell your cousin that the position of the table is about 10 m from the statue, toward the slide.

✔️ **Reading Check** How could you describe the position of a different table using the statue as a reference point?

## Changing the Reference Point

The description of an object's position depends on the reference point. Suppose you choose the drinking fountain in **Figure 1** as the reference point instead of the statue. You could say that the direction of the table is toward the dead tree. Now the distance is measured from the drinking fountain to the table. You could tell your cousin that the table is about 12 m from the drinking fountain, toward the dead tree. The description of the table's position changed because the reference point is different. Its actual position did not change at all.

✔️ **Key Concept Check** How does the description of an object's position depend on a reference point?

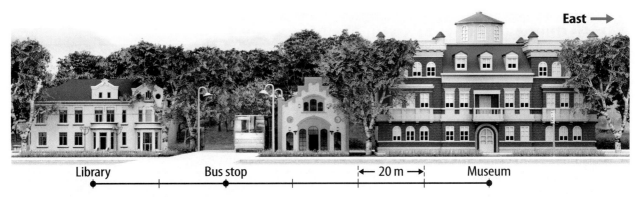

East →

Library        Bus stop        |← 20 m →|        Museum

## The Reference Direction

When you describe an object's position, you compare its location to a reference direction. In **Figure 1,** the reference direction for the first reference point is toward the slide. Sometimes the words *positive* and *negative* are used to describe direction. The reference direction is the positive (+) direction. The opposite direction is the negative (−) direction. Suppose you **specify** east as the reference direction in **Figure 2.** You could say the museum's entrance is +80 m from the bus stop. The library's entrance is −40 m from the bus stop. To a friend, you would probably just say the museum is two buildings east of the bus stop and the library is the building west of the bus stop. Sometimes, however, using the words *positive* and *negative* to describe direction is useful for explaining changes in an object's position.

**Figure 2** If east is the reference direction, then the museum is in the positive direction from the bus stop. The library is in the negative direction.

**ACADEMIC VOCABULARY**

specify
*(verb)* to indicate or identify

---

Inquiry **MiniLab**                                    **10 minutes**

### Why is a reference point useful?

To find an object's position, you need to know its distance and direction from a reference point.

1. Read and complete a lab safety form.

2. Put a **sticky note** at the 50-cm mark of a **meterstick.** This is your reference point.

3. Place a **small object** at the 40-cm mark. It is 10 cm in the negative direction from the reference point.

4. Copy the table in your Science Journal. Continue moving the object and recording its distance, its reference direction, and its position to complete the table.

| Position of Object | | |
|---|---|---|
| Distance (cm) | Reference Direction | Position (cm) |
| 10 cm | negative | 40 cm |
| 40 cm | positive | |
| 15 cm | positive | |
| | positive | 75 cm |
| | | 30 cm |

### Analyze and Conclude

1. **Recognize Cause and Effect** How would the data in the table change if the positions were the same but the reference point was at the 40-cm mark?

2. 🔑 **Key Concept** Why is a reference point useful in describing positions of an object?

**Figure 3** You need two reference directions to describe the position of a building in the city.

Your classmate's home

Your house is the reference point to find your classmate's house.

Library

✓ **Visual Check** If the library is the reference point, how would you describe the position of your house in two dimensions?

REVIEW VOCABULARY ·····

**dimension**
distance or length measured in one direction

## Describing Position in Two Dimensions

You were able to describe the position of the picnic table in the park in one **dimension.** Your cousin had to walk in only one direction to reach the table. But sometimes you need to describe an object's position using more than one reference direction. The city shown in **Figure 3** is an example. To describe the position of a house in a city might require two reference directions. When you describe a position using two directions, you are using two dimensions.

### Reference Directions in Two Dimensions

To describe a position on the map in **Figure 3,** you might choose north and east or south and west as reference directions. Sometimes north, south, east, and west are not the most useful reference directions. If you are playing checkers and want to describe the position of a certain checker, you might use "right" and "forward" as reference directions. If you are describing the position of a certain window on a skyscraper, you might choose "left" and "up" as reference directions.

✓ **Reading Check** What are two other reference directions you might use to describe the position of a building in a city?

### Locating a Position in Two Dimensions

Finding a position in two dimensions is similar to finding a position in one dimension. First, choose a reference point. To locate your classmate's home on the map in **Figure 3,** you could use your home as a reference point. Next, specify reference directions—north and east. Then, determine the distance along each reference direction. In the figure, your classmate's house is one block north and two blocks east of your house.

✓ **Key Concept Check** How can you describe the position of an object in two dimensions?

# Describing Changes in Position

Sometimes you need to describe how an object's position changes. You can tell that the boat in **Figure 4** moved because its position changed relative, or compared, to the buoy. **Motion** *is the process of changing position.*

## Motion Relative to a Reference Point

Is the man in the boat in **Figure 4** in motion? Suppose the fishing pole is the reference point. Because the positions of the man and the pole do not change relative to each other, the man does not move relative to the pole. Now suppose the buoy is the reference point. Because the man's distance from the buoy changes, he is in motion relative to the buoy.

**WORD ORIGIN**

**motion**
from Latin *motere*; means "to move"

◀ **Figure 4** The man in the boat is not in motion compared to his fishing pole. He is in motion compared to the buoy.

## Distance and Displacement

Suppose a baseball player runs the bases, as shown in **Figure 5**. Distance is the length of the path the player runs, as shown by the red arrows. **Displacement** *is the difference between the initial (first) position and the final position of an object.* It is shown in the figure by the blue arrows. Notice that distance and displacement are equal only if the motion is in one direction.

 **Key Concept Check** What is the difference between distance and displacement?

**Figure 5** Distance depends on the path taken. Displacement depends only on the initial and final positions. ▼

## Distance and Displacement 🗝

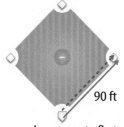

90 ft

When a player runs to first base, the distance is 90 ft and the displacement is 90 ft toward first base.

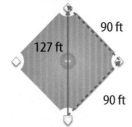

90 ft
127 ft
90 ft

When a player runs to second base, the distance is 180 ft, but the displacement is 127 ft toward second base.

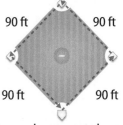

90 ft   90 ft
90 ft   90 ft

When a player runs to home base, the distance is 360 ft, but the displacement is 0 ft.

# Lesson 1 Review

## Visual Summary

A reference point, a reference direction, and distance are needed to describe the position of an object.

An object is in motion if its position changes relative to a reference point.

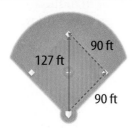

The distance an object moves and the object's displacement are not always the same.

**FOLDABLES**

Use your lesson Foldable to review the lesson. Save your Foldable for the project at the end of the chapter.

## What do you think NOW?

You first read the statements below at the beginning of the chapter.

**1.** Displacement is the distance an object moves along a path.

**2.** The description of an object's position depends on the reference point.

Did you change your mind about whether you agree or disagree with the statements? Rewrite any false statements to make them true.

## Use Vocabulary

**1** **Define** *motion* in your own words.

**2** The difference between the initial position and the final position of an object is its _____.

## Understand Key Concepts

**3** **Explain** why a description of position depends on a reference point.

**4** To describe a position in more than one dimension, you must use more than one
   **A.** displacement.      **C.** reference point.
   **B.** reference direction.  **D.** type of motion.

**5** **Apply** If you walk 2 km from your house to a store and then back home, what is your displacement?

## Interpret Graphics

**6** **Interpret** Using 12 as the reference point, how you can tell that the hands of the clock on the right have moved from their previous position, shown on the left?

**7** **Summarize** Copy and fill in the graphic organizer below to identify the three things that must be included in the description of position.

## Critical Thinking

**8** **Compare** Relative to some reference points, your nose is in motion when you run. Relative to others, it is not in motion. Give one example of each.

# GPS to the Rescue!

*How Technology Helps Bring Home Family Pets*

Satellite

Cell phone tracking display ▶

**Y**ou've seen the signs tacked to streetlights and telephone poles: *LOST! Golden retriever. Reward. Please Call!* Losing a pet can be heartbreaking. Fortunately, there's an alternative to posting fliers—a pet collar with a Global Positioning System (GPS) chip that helps locate the pet. Here is how GPS can help you track or locate your pet:

**1** **GPS is a network of at least 24 satellites in orbit around Earth. Each satellite circles Earth twice a day and sends information to ground receivers.**

Cell phone tower

GPS Collar

**4** **A GPS pet collar works much the same as any other GPS receiver. Once it is activated, the collar can transmit a message to a Web site or to the owner's cell phone.**

**2** **GPS satellites act as reference points. Ground-based GPS receivers compare the time a signal is transmitted by a satellite to the time it is received on Earth. The difference indicates the satellite's distance. Signals from as many as four satellites are used to pinpoint a user's exact position.**

**3** **GPS uses computer technology to calculate location, speed, direction, and time. The same GPS technology used to locate or guide airplanes, cars, and campers can help find a lost pet anywhere on Earth!**

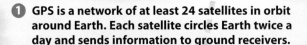

## It's Your Turn

**DESIGN** GPS technology has revolutionized the way people track and locate almost everything. Can you think of a new application for GPS technology? Write an advertisement or a TV commercial for a new idea that puts GPS technology to work!

# Speed and Velocity

### Reading Guide

**Key Concepts**
**ESSENTIAL QUESTIONS**

- What is speed?
- How can you use a distance-time graph to calculate average speed?
- What are ways velocity can change?

**Vocabulary**

**speed** p. 17

**constant speed** p. 18

**instantaneous speed** p. 18

**average speed** p. 19

**velocity** p. 23

g **Multilingual eGlossary**

### Inquiry How Fast?

When you hear the word *cheetah,* you might think of how fast a cheetah can run. As the fastest land animal on Earth, it can reach a speed of 30 m/s for a short period of time. Other than how fast it runs, how might you describe the motion of a cheetah?

## inquiry Launch Lab

### How can motion change?

Have you ever used a tube slide at a playground or at a water park? You can build a marble tube slide from foam tubes. You can then use the slide to observe how the motion of a marble changes as it rolls down the slide.

1. Read and complete a lab safety form.

2. Join **foam tubes** into a tube slide, using **masking tape** to connect the tubes. Use a desk and other objects to support the slide so that it changes direction and slopes toward the floor.

3. Drop a **marble** into the top of the slide. Observe the changes in its motion as it rolls through the tubes. Have the marble roll into a **container** at the bottom of the slide.

### Think About This

1. In what ways did the motion of the marble change as it moved down the slide?

2. 🔑 **Key Concept** At what parts of the tube slide did the marble move fastest? At what parts did it move slowest?

## What is speed?

How fast do you walk when you are hungry and there is good food on the table? How fast do you move when you have a chore to do? Sometimes you move quickly, and sometimes you move slowly. One way you can describe how fast you move is to determine your speed. **Speed** *is a measure of the distance an object travels per unit of time.*

🔑 **Key Concept Check** What is speed?

### Units of Speed

You can calculate speed by dividing the distance traveled by the time it takes to go that distance. The units of speed are units of distance divided by units of time. The SI unit for speed is meters per second (m/s). Other units are shown in **Table 1.** What units of distance and time are used in each example?

| Table 1  Typical Speeds 🔑 | |
|---|---|
| **Airplane** 245 m/s 882 km/h 548 mph | 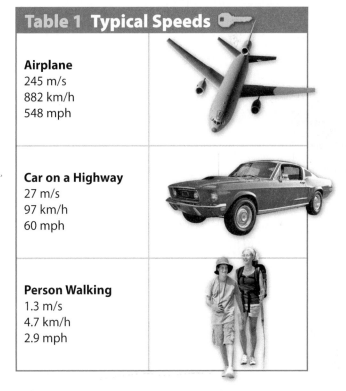 |
| **Car on a Highway** 27 m/s 97 km/h 60 mph | |
| **Person Walking** 1.3 m/s 4.7 km/h 2.9 mph | |

**Table 1** Different units of distance and time can be used to determine units of speed.

## Constant and Changing Speed

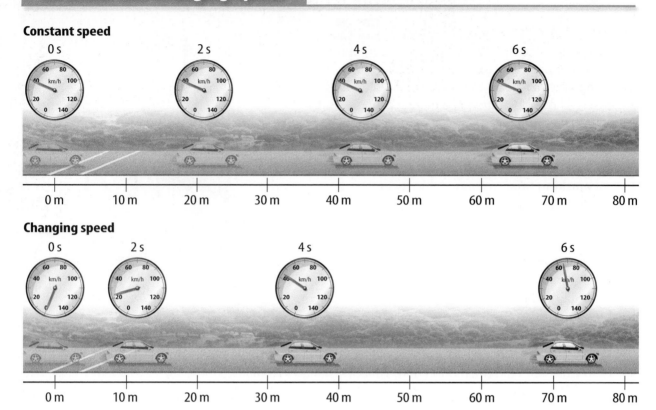

**Constant speed**

**Changing speed**

**Figure 6** When the car moves with constant speed, it moves the same distance each period of time. When the car's speed changes, it moves a different distance each period of time.

✅ **Visual Check** What is the bottom car's instantaneous speed at 6 s?

## Constant Speed

What happens to your speed when you ride in a car? Sometimes you ride at a steady speed. If you move away from a stop sign, your speed increases. You slow down when you pull into a parking space.

Think about a time that a car's speed does not change. As the car at the top of **Figure 6** travels along the road, the speedometer above each position shows that the car is moving at the same speed at each location and time. Each second, the car moves 11 m. Because it moves the same distance each second, its speed is not changing. **Constant speed** *is the rate of change of position in which the same distance is traveled each second.* The car is moving at a constant speed of 11 m/s.

## Changing Speed

How is the motion of the car at the bottom of **Figure 6** different from the motion of the car at the top? Between 0 s and 2 s, the car at the bottom travels about 10 m. Between 4 s and 6 s, however, the car travels more than 20 m. Because the car travels a different distance each second, its speed is changing.

If the speed of an object is not constant, you might want to know its speed at a certain moment. **Instantaneous speed** *is speed at a specific instant in time.* You can see a car's instantaneous speed on its speedometer.

✅ **Reading Check** How would the distance the car travels each second change if it were slowing down?

## Average Speed

Describing an object's speed is easy if the speed is constant. But how can you describe the speed of an object when it is speeding up or slowing down? One way is to calculate its average speed. **Average speed** *is the total distance traveled divided by the total time taken to travel that distance.* You can calculate average speed using the equation below.

### Average Speed Equation

average speed (in m/s) $= \dfrac{\text{total distance (in m)}}{\text{total time (in s)}}$

$$\bar{v} = \dfrac{d}{t}$$

The symbol $\bar{v}$ represents the term "average velocity." You will read more about velocity, and how it relates to speed, later in this lesson. However, at this point, $\bar{v}$ is simply used as the symbol for "average speed." The SI unit for speed, meters per second (m/s), is used in the above equation. You could instead use other units of distance and time in the average speed equation, such as kilometers and hours.

### Math Skills ✖️➗➕  Average Speed Equation

**Solve for Average Speed** Melissa shot a model rocket 360 m into the air. It took the rocket 4 s to fly that far. What was the average speed of the rocket?

**1** **This is what you know:**
distance: $d = 360$ m
time: $t = 4$ s

**2** **This is what you need to find:** average speed: $\bar{v}$

**3** **Use this formula:** $\bar{v} = \dfrac{d}{t}$

**4** **Substitute:** $\bar{v} = \dfrac{360 \text{ m}}{4 \text{ s}} = 90$ m/s
the values for *d* and *t* into the formula and divide.

**Answer:** The average speed was **90 m/s.**

- Math Practice
- Personal Tutor

### Practice

1. It takes Ahmed 50 s on his bicycle to reach his friend's house 250 m away. What is his average speed?

2. A truck driver makes a trip that covers 2,380 km in 28 hours. What is the driver's average speed?

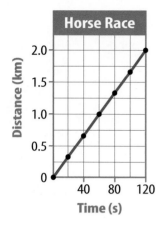

**Horse Race**

_Distance (km)_ vs _Time (s)_

**Figure 7** According to this distance-time graph, it took 2 min for the horse to run 2 km.

 **Visual Check** How would this distance-time graph be different if the horse's speed changed over time?

# Distance-Time Graphs

The Kentucky Derby is often described as the most exciting two minutes in sports. The thoroughbred horses in this race run for a distance of 2 km. The speeds of the horses change many times during a race, but **Figure 7** describes what a horse's motion might be if its speed did not change. The graph shows the distance a horse might travel when distance measurements are made every 20 seconds. Follow the height of the line from the left side of the graph to the right side. You can see how the distance the horse ran changed over time.

Graphs like the one in **Figure 7** can show how one measurement compares to another. When you study motion, two measurements frequently compared to each other are distance and time. The graphs that show these comparisons are called distance-time graphs. Notice that the change in the distance the horse ran around the track is the same each second on the graph. This means the horse was moving with a constant speed. Constant speed is shown as a straight line on a distance-time graph.

**Reading Check** How is constant speed shown on a distance-time graph?

---

**Inquiry MiniLab** 15 minutes

## How can you graph motion?

You can represent motion with a distance-time graph.

1. Read and complete a lab safety form.
2. Use **masking tape** to mark a starting point on the floor.
3. As you cross the starting point, start a **stopwatch.** Stop walking after 2 s. Measure the distance with a **meterstick.** Record the time and distance in your Science Journal.
4. Repeat step 3 by walking at about the same speed for 4 s and then for 6 s.
5. Use the graph in **Figure 7** as an example to create a distance-time graph of your data. The line on the graph should be as close to the points as possible.

### Analyze and Conclude

1. **Predict** Based on the graph, how far would you probably walk at the same speed in 8 s?

2. **Key Concept** Look back at the average speed equation. Explain how you could use your graph to find your average walking speed.

## Comparing Speeds on a Distance-Time Graph

You can use distance-time graphs to compare the motion of two different objects. **Figure 8** is a distance-time graph that compares the motion of two horses that ran the Kentucky Derby. The motion of horse A is shown by the blue line. The motion of horse B is shown by the orange line. Look at the far right side of the graph. When horse A reached the finish line, horse B was only 1.5 km from the starting point of the race.

Recall that average speed is distance traveled divided by time. Horse A traveled a greater distance than horse B in the same amount of time. Horse A had greater average speed. Compare how steep the lines are on the graph. The measure of steepness is the slope. The steeper the line, the greater the slope. The blue line is steeper than the orange line. Steeper lines on distance-time graphs indicate faster speeds.

## Using a Distance-Time Graph to Calculate Speed

You can use distance-time graphs to calculate the average speed of an object. The motion of a trail horse traveling at a constant speed is shown on the graph in **Figure 9.** The steps needed to calculate the average speed from a distance-time graph also are shown in the figure.

 **Key Concept Check** How can you use a distance-time graph to calculate average speed?

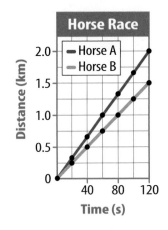

**Figure 8** You can tell that horse A ran a faster race than horse B because the blue line is steeper than the orange line.

---

**Average Speed** 🔑

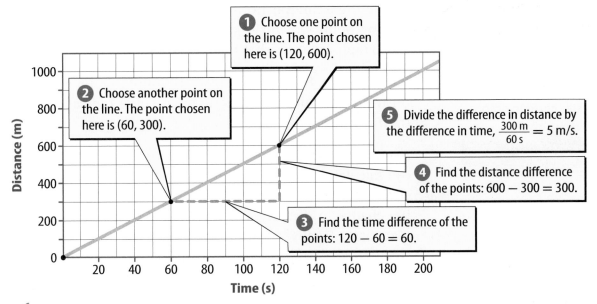

**Figure 9** The average speed of the horse from 60 s to 120 s can be calculated from this distance-time graph.

❶ Choose one point on the line. The point chosen here is (120, 600).

❷ Choose another point on the line. The point chosen here is (60, 300).

❸ Find the time difference of the points: 120 − 60 = 60.

❹ Find the distance difference of the points: 600 − 300 = 300.

❺ Divide the difference in distance by the difference in time, $\frac{300\ m}{60\ s} = 5\ m/s$.

**Visual Check** How does the average speed of the horse from 60 s to 120 s compare to its average speed from 120 s to 180 s?

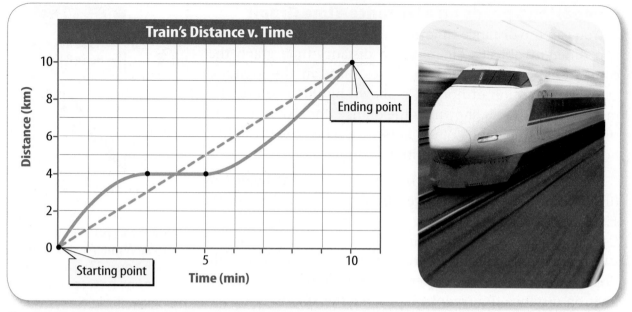

**Train's Distance v. Time**

Starting point

Ending point

**Figure 10** Even though the train's speed is not constant, you can calculate its average speed from a distance-time graph.

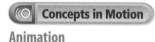
**Concepts in Motion**

**Animation**

Make a horizontal three-tab book and label it as shown. Use it to summarize how changes in speed are represented on a distance-time graph.

Slowing Down | Stopping | Speeding Up

## Distance-Time Graph and Changing Speed

So far, the distance-time graphs in this lesson have included straight lines. Distance-time graphs have straight lines only for objects that move at a constant speed. The graph in **Figure 10** for the motion of a train is different. Because the speed of the train changes instead of being constant, its motion on a distance-time graph is a curved line.

**Slowing Down** Notice how the shape of the line in **Figure 10** changes. Between 0 min and 3 min, its slope decreases. The downward curve indicates that the train slowed down.

**Stopping** What happened between 3 min and 5 min? The line during these times is horizontal. The train's distance from the starting point remains 4 km. A horizontal line on a distance-time graph indicates that there is no motion.

**Speeding Up** Between 5 min and 10 min, the slope of the line on the graph increases. The upward curve indicates that the train was speeding up.

**Average Speed** Even when the speed of an object changes, you can calculate its average speed from a distance-time graph. First, choose a starting point and an ending point. Next, determine the change in distance and the change in time between these two points. Finally, substitute these values into the average speed equation. The slope of the dashed line in **Figure 10** represents the train's average speed between 0 minutes and 10 minutes.

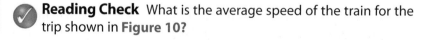
**Reading Check** What is the average speed of the train for the trip shown in **Figure 10**?

# Velocity

Often, describing just the speed of a moving object does not completely describe its motion. If you describe the motion of a bouncing ball, for example, you would also describe the direction of the ball's movement. Both speed and direction are part of motion. **Velocity** *is the speed and the direction of a moving object.*

## Representing Velocity

In Lesson 1, an arrow represented the displacement of an object from a reference point. The velocity of an object also can be represented by an arrow, as shown in **Figure 11.** The length of the arrow indicates the speed. A greater speed is shown by a longer arrow. The arrow points in the direction of the object's motion.

In **Figure 11,** both students are walking at 1.5 m/s. Because the speeds are equal, both arrows are the same length. But the girl is walking to the left and the boy is walking to the right. The arrows point in different directions. The students have different velocities because each student has a different direction of motion.

## Changes in Velocity

Look at the bouncing ball in **Figure 12.** Notice how from one position to the next, the arrows showing the velocity of the ball change direction and length. The changes in the arrows mean that the velocity is constantly changing. Velocity changes when the speed of an object changes, when the direction that the object moves changes, or when both the speed and the direction change. You will read about changes in velocity in Lesson 3.

 **Key Concept Check** How can velocity change?

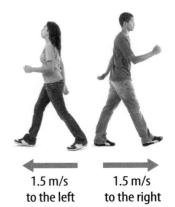

1.5 m/s to the left    1.5 m/s to the right

▲ **Figure 11** The students are walking with the same speed but different velocities.

**WORD ORIGIN** · · · · · · · · · ·

**velocity**
from Latin *velocitas;* means "swiftness, speed"

**Figure 12** The velocity of the ball changes continually because both the speed and the direction of the ball change as the ball bounces. ▼

## Changing Speed and Direction 🗝

Velocity

**Visual Check** Are there two positions of the bouncing ball in which the velocity is the same? Explain.

# Lesson 2 Review

## Visual Summary

Speed is a measure of the distance an object travels per unit of time. You can describe an object's constant speed, instantaneous speed, or average speed.

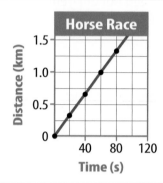

A distance-time graph shows the speed of an object.

Velocity includes both the speed and the direction of motion.

**FOLDABLES**

Use your lesson Foldable to review the lesson. Save your Foldable for the project at the end of the chapter.

## What do you think NOW?

You first read the statements below at the beginning of the chapter.

**3.** Constant speed is the same thing as average speed.

**4.** Velocity is another name for speed.

Did you change your mind about whether you agree or disagree with the statements? Rewrite any false statements to make them true.

## Use Vocabulary

**1** **Distinguish** between speed and velocity.

**2** **Define** *constant speed* in your own words.

## Understand Key Concepts

**3** **Recall** How can you calculate average speed from a distance-time graph?

**4** **Analyze** Describe three ways a bicyclist can change velocity.

**5** Which choice is a unit of speed?
- **A.** h/mi
- **B.** km/h
- **C.** $m^2/s$
- **D.** $N \cdot m^2$

## Interpret Graphics

**6** **Organize Information** Copy and fill in the graphic organizer below to show possible steps for making a distance-time graph.

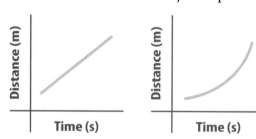

**7** **Interpret** What does the shape of each line indicate about the object's speed?

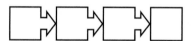

## Critical Thinking

**8** **Decide** Aaron leaves one city at noon. He has to be at another city 186 km away at 3:00 P.M. The speed limit the entire way is 65 km/h. Can he arrive at the second city on time? Explain.

## Math Skills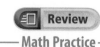
— Math Practice —

**9** A train traveled 350 km in 2.5 h. What was the average speed of the train?

# What do you measure to calculate speed?

## Materials

meterstick

stopwatch

wind-up toys (4)

calculator

graph paper

**Safety**

You turn on the television and see a news report. It shows trees that are bent almost to the ground because of a strong wind. Is it a hurricane or a tropical storm? The type of storm depends on the speed of the wind. A meteorologist must measure both distance and time before calculating the wind's speed.

## Learn It

When you **measure,** you use a tool to find a quantity. To find the average speed of an object, you measure the distance it travels and the time it travels. You can then calculate speed using the average speed equation. In this lab, you use distance and time measurements to calculate speeds of moving toys.

## Try It

1 Read and complete a lab safety form.

2 Copy the data table on this page into your Science Journal. Add more lines as you need them.

3 Choose appropriate starting and ending points on the floor. Use a meterstick to measure the distance between these points. Record this distance to the nearest centimeter.

4 Wind one toy. Measure in tenths of a second the time the toy takes to travel from start to finish. Record the time in the data table.

5 Repeat steps 3 and 4 for three more toys. Vary the distance from start to finish for each toy.

## Apply It

6 **Calculate** the average speed of each toy. Record the speeds in your data table.

7 **Create** a bar graph of your data. Place the name of each toy on the *x*-axis and the average speed on the *y*-axis.

8 🔑 **Key Concept** Use the definition of *speed* to explain why the average speeds of the toys can be compared, even though the toys traveled different distances.

| Toy Speeds | | | |
|---|---|---|---|
| Toy | Distance (m) | Time (s) | Average Speed (m/s) |
| | | | |
| | | | |
| | | | |
| | | | |

# Lesson 3

## Reading Guide

### Key Concepts
ESSENTIAL QUESTIONS

- What are three ways an object can accelerate?

- What does a speed-time graph indicate about an object's motion?

### Vocabulary

**acceleration** p. 27

 **Multilingual eGlossary**

**Video** BrainPOP®

# Acceleration

## Inquiry Is velocity changing?

How does the velocity of this motorcycle racer change as he speeds along the track? As he enters a curve, he slows down, leans to the side, and changes direction. On a straightaway, he speeds up and moves in a straight line. How can the velocity of a moving object change?

## In what ways can velocity change?

As you walk, your motion changes in many ways. You probably slow down when the ground is uneven. You might speed up when you realize that you are late for dinner. You change direction many times. What would these changes in velocity look like on a distance-time graph?

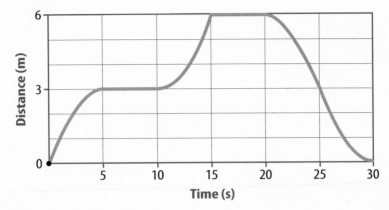

1. Read and complete a lab safety form.

2. Use a **meterstick** to measure a 6-m straight path along the floor. Place a mark with **masking tape** at 0 m, 3 m, and 6 m.

3. Look at the graph above. Decide what type of motion occurs during each 5-second period.

4. Try to walk along your path according to the motion shown on the graph. Have your partner time your walk with a **stopwatch.** Switch roles, and repeat this step.

### Think About This

1. What does a horizontal line segment on a distance-time graph indicate?

2. **Key Concept** According to the graph, at what times do the following motions take place? **a.** You change direction. **b.** Your speed increases. **c.** Your speed decreases.

## Acceleration—Changes in Velocity

Imagine riding in a car. The driver steps on the gas pedal, and the car moves faster. Moving faster means the car's velocity increases. The driver then takes her foot off the pedal, and the car's velocity decreases. Next the driver turns the steering wheel. The car's velocity changes because its direction changes. The car's velocity changes if either the speed or the direction of the car changes.

When a car's velocity changes, the car is accelerating. **Acceleration** *is a measure of the change in velocity during a period of time.* An object accelerates when its velocity changes as a result of increasing speed, decreasing speed, or changing direction.

You might have experienced a large acceleration if you have ever ridden a roller coaster. Think about all the changes in speed and direction you experience on a roller coaster ride. When you drop down a hill of a roller coaster, you reach a faster speed quickly. The roller coaster is accelerating because its speed is increasing. The roller coaster also accelerates any time it changes direction. It accelerates again when it slows down and stops at the end of the ride. Each time the velocity of the roller coaster changes, it accelerates.

✔ **Reading Check** What is acceleration?

## Ways an Object Can Accelerate 🔑

Figure 13 Acceleration occurs when an object speeds up, slows down, or changes its direction of motion.

**Speeding Up**

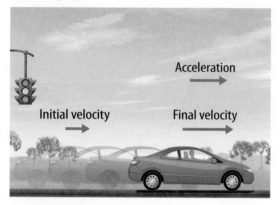

Acceleration

Initial velocity    Final velocity

**Slowing Down**

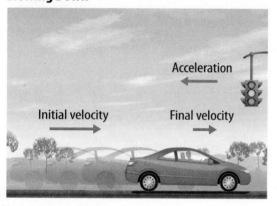

Acceleration

Initial velocity    Final velocity

**Changing Direction**

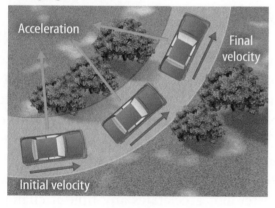

Acceleration

Final velocity

Initial velocity

✅ **Visual Check** If the car in the top picture moved faster, how would the acceleration arrow change?

## Representing Acceleration

Like velocity, acceleration has a direction and can be represented by an arrow. Ways an object can accelerate are shown in **Figure 13**. The length of each blue acceleration arrow indicates the amount of acceleration. An acceleration arrow's direction depends on whether velocity increases or decreases.

## Changing Speed

The car in the top picture of **Figure 13** is speeding up. At first it is moving slowly, so the arrow that represents its initial velocity is short. The car's speed increases, so the final velocity arrow is longer. As velocity increases, the car accelerates. Notice that the acceleration arrow points in the same direction as the velocity arrows.

The car in the middle picture of **Figure 13** is slowing down. At first it moves fast, so the arrow showing its velocity is long. After the car slows down, the arrow showing its final velocity is shorter. When velocity decreases, acceleration and velocity are in opposite directions. The arrow that represents acceleration is pointing in the direction opposite to the direction the car is moving.

✅ **Reading Check** In what direction is acceleration if an object is slowing down?

## Changing Direction

The car in the bottom picture of **Figure 13** has a constant speed, so the velocity arrow is the same length at each point in the turn. But the car's velocity changes because its direction changes. Because velocity changes, the car is accelerating. Notice the direction of the blue acceleration arrows. It might surprise you that the car is accelerating toward the inside of the curve. Recall that acceleration is the change in velocity. If you compare one velocity arrow with the next, you can see that the change is always toward the inside of the curve.

🔑 **Key Concept Check** What are three ways an object can accelerate?

# Calculating Acceleration

Acceleration is a change in velocity divided by the time interval during which the velocity changes. Recall that "velocity" is the speed of an object in a given direction. However, if an object moves along a straight line, you can calculate its acceleration without considering the object's direction. In this lesson, "velocity" refers to only an object's speed. Positive acceleration can be thought of as speeding up in the forward direction. Negative acceleration is slowing down in the forward direction as well as speeding up in the reverse direction.

## Acceleration Equation

acceleration (in m/s²) =

$$\frac{\text{final speed (in m/s)} - \text{initial speed (in m/s)}}{\text{total time (in s)}}$$

$$a = \frac{v_f - v_i}{t}$$

Acceleration has SI units of meters per second per second (m/s/s). This can also be written as meters per second squared (m/s²).

## Math Skills  Acceleration Equation

**Solve for Acceleration** A bicyclist started from rest along a straight path. After 2.0 s, his speed was 2.0 m/s. After 5.0 s, his speed was 8.0 m/s. What was his acceleration during the time 2.0 s to 5.0 s?

**1 This is what you know:**

initial speed: $v_i$ = 2.0 m/s

final speed: $v_f$ = 8.0 m/s

total time: $t$ = 5.0 s − 2.0 s = 3.0 s

**2 This is what you need to find:** acceleration: $a$

**3 Use this formula:** $a = \dfrac{v_f - v_i}{t}$

**4 Substitute:** the values for $v_i$, $v_f$, and $t$ into the formula; subtract; then divide.

$$a = \frac{8.0 \text{ m/s} - 2.0 \text{ m/s}}{3.0 \text{ s}} = \frac{6.0 \text{ m/s}}{3.0 \text{ s}} = 2.0 \text{ m/s}^2$$

**Answer:** The acceleration of the bicyclist was **2.0 m/s².**

• **Math Practice**
• **Personal Tutor**

## Practice

Aidan drops a rock from a cliff. After 4.0 s, the rock is moving at 39.2 m/s. What is the acceleration of the rock?

## inquiry MiniLab

**10 minutes**

### How is a change in speed related to acceleration?

What happens if the distance you walk each second increases? Follow these steps to demonstrate acceleration.

1. Read and complete a lab safety form.

2. Use **masking tape** to mark a course on the floor. Mark start, and place marks along a straight path at 10 cm, 40 cm, 90 cm, 160 cm, and 250 cm from the start.

3. Clap a steady beat. On the first beat, the person walking the course is at start. On the second beat, the walker should be at the 10-cm mark, and so on.

### Analyze and Conclude

1. **Explain** what happened to your speed as you moved along the course.

2. **Key Concept** Suppose your speed at the final mark was 0.95 m/s. Calculate your average acceleration from start through the final segment of the course.

---

**WORD ORIGIN** ············

**horizontal**
from Greek *horizein*, means "limit, divide, separate"

**vertical**
from Latin *verticalis*, means "overhead"

## Speed-Time Graphs

Recall that you can show an object's speed using a distance-time graph. You also can use a speed-time graph to show how speed changes over time. Just like a distance-time graph, a speed-time graph has time on the **horizontal** axis—the *x*-axis. But speed is on the **vertical** axis—the *y*-axis. The figures on the next few pages compare distance-time graphs and speed-time graphs for different types of motion.

### Object at Rest

An object at rest is not moving, so its speed is always zero. As a result, the speed-time graph for an object at rest is a horizontal line at *y* = 0, as shown in **Figure 14.**

**Figure 14** Both the distance-time graph and the speed-time graph are horizontal lines for an object at rest.

**Distance**

**Time**

The object's distance from the reference point does not change.

**Speed**

**Time**

The speed is zero and does not change.

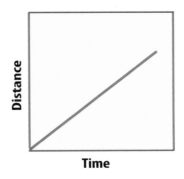

The distance increases at a steady rate over time.

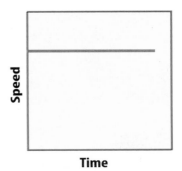

The object's speed does not change.

## Constant Speed

Think about a farm machine moving through a field at a constant speed. At every point in time, its speed is the same. If you plot its speed on a speed-time graph, the plotted line is horizontal, as shown in **Figure 15.** The speed of the object is represented by the distance the horizontal line is from the *x*-axis. If the line is farther from the *x*-axis, the object is moving at a faster speed.

## Speeding Up

A plane speeds up as it moves down a runway and takes off. Suppose the speed of the plane increases at a steady rate. If you plot the speed of the plane on a speed-time graph, the line might look like the one in **Figure 16.** The line on the speed-time graph is closer to the *x*-axis at the beginning of the time period when the plane has a lower speed. It slants upward toward the right side of the graph as the speed increases.

 **Reading Check** Why does the speed-time graph of an object that is speeding up slope upward from left to right?

▲ **Figure 15** For an object moving at constant speed, the speed-time graph is a horizontal line.

**Figure 16** The line on the speed-time graph for an object that is speeding up has an upward slope. ▼

As the distance increases, the rate of increase gets larger over time.

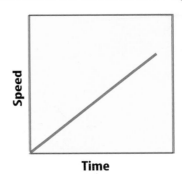

The speed of the object increases at a steady rate over time.

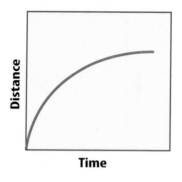

As the distance increases, the rate of increase gets smaller over time.

The speed of the object decreases at a steady rate over time.

▲ **Figure 17**  The line on the speed-time graph for an object that is slowing down has a downward slope.

### Slowing Down

The speed-time graph in **Figure 17** shows the motion of a space shuttle just after it lands. It slows down at a steady rate and then stops. Initially, the shuttle is moving at a high speed. The point representing this speed is far from the *x*-axis. As the shuttle's speed decreases, the points representing its speed are closer to the *x*-axis. The line on the speed-time graph slopes downward to the right. When the line touches the *x*-axis, the speed is zero and the shuttle is stopped.

**Key Concept Check** What does a speed-time graph show about the motion of an object?

### Limits of Speed-Time Graphs

You have read that distance-time graphs show the speed of an object. However, they do not describe the direction in which an object is moving. In the same way, speed-time graphs show only the relationship between speed and time. A speed-time graph of the skier in **Figure 18** would show changes in his speed. It would not show what happens when the skier's velocity changes as the result of a change in his direction.

**Figure 18** A speed-time graph of the motion of this skier would show changes in speed but not changes in direction. ▼

## Summarizing Motion

Now that you know about motion, how might you describe a walk down the hallway at school? You can describe your position by your direction and distance from a reference point. You can compare your distance and your displacement and find your average speed. You know that you have an instantaneous speed and can tell when you walk at a constant speed. You can describe your velocity by your speed and your direction. You know you are accelerating if your velocity is changing.

**Visual Check** The skier slows down and speeds up along the curved path. Describe a speed-time graph of this motion.

## Visual Summary

An object accelerates if it speeds up, slows down, or changes direction.

Acceleration in a straight line can be calculated by dividing the change in speed by the change in time.

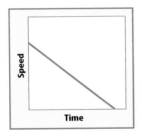

Initial velocity          Final velocity

A speed-time graph shows how an object's speed changes over time.

**FOLDABLES**

Use your lesson Foldable to review the lesson. Save your Foldable for the project at the end of the chapter.

## What do you think NOW?

You first read the statements below at the beginning of the chapter.

**5.** You can calculate acceleration by dividing the change in velocity by the change in distance.

**6.** An object accelerates when either its speed or its direction changes.

Did you change your mind about whether you agree or disagree with the statements? Rewrite any false statements to make them true.

## Use Vocabulary

**1** **Define** *acceleration* in your own words.

**2** **Use the term** *acceleration* in a complete sentence.

## Understand Key Concepts

**3** **Recall** how a roller coaster can accelerate, even when it is moving at a constant speed.

**4** A speed-time graph is a horizontal line with a *y*-value of 4. Which describes the object's motion?

   **A.** at rest       **C.** slowing down

   **B.** constant speed    **D.** speeding up

## Interpret Graphics

**5** **Organize Information** Copy and fill in the graphic organizer below for the four types of speed-time graphs. For each, describe the motion of the object.

## Critical Thinking

**6** **Evaluate** A race car accelerates on a straight track from 0 to 100 km/h in 6 s. Another race car accelerates from 0 to 100 km/h in 5 s. Compare the velocities and accelerations of the cars during their races.

## Math Skills ✕∸∻     Review

— Math Practice —

**7** After 2.0 s, Isabela was riding her bicycle at 3.0 m/s on a straight path. After 5.0 s, she was moving at 5.4 m/s. What was her acceleration?

**8** After 3.0 s, Mohammed was running at 1.2 m/s on a straight path. After 7.0 s, he was running at 2.0 m/s. What was his acceleration?

## Materials

metersticks (6)

stopwatches
(6)

masking tape

tennis ball

## Safety

# Calculate Average Speed from a Graph

You probably do not walk the same speed uphill and downhill, or when you are just starting out and when you are tired. If you are walking and you measure and record the distance you walk every minute, the distances will vary. How might you use these measurements to calculate the average speed you walked? One way is to organize the data on a distance-time graph. In this activity, you will use such a graph to compare average speeds of a ball on a track using different heights of a ramp.

## Ask a Question

How does the height of a ramp affect the speed of a ball along a track?

## Make Observations

**1** Read and complete a lab safety form.

**2** Make a 3-m track. Place three metersticks end-to-end. Place three other metersticks end-to-end about 6 cm from the first set of metersticks. Use tape to hold the metersticks in place. Mark each half-meter with tape. Use books to make a ramp leading to the track.

**3** A student should be at each half-meter mark with a stopwatch. Another student should be by the ramp to roll a ball along the track.

**4** When the ball passes start, all group members should start their stopwatches. Each student should stop his or her stopwatch when the ball crosses the mark where the student is stationed.

**5** Practice several times to get consistent rolls and times.

## Form a Hypothesis

**6** Create a hypothesis about how the number of books used as a ramp affects the speed of the ball rolling along the track.

## Test Your Hypothesis

**7** Write a plan for varying the number of books and making distance and time measurements.

**8** Create a data table in your Science Journal that matches your plan. A sample is shown to the right.

**9** Use your plan to make the measurements. Record them in the data table.

**10** Plot the data for each height of the ramp on a graph that shows the distance the ball traveled on the *x*-axis and time on the *y*-axis. For each ramp height, draw a straight line that goes through the most points.

**11** Choose two points on each line. Calculate the average speed between these points by dividing the difference in the distances for the two points by the difference in the times.

## Analyze and Conclude

**12** **Compare** the average speeds for each ramp height. Use this comparison to decide whether your results support your hypothesis.

**13** **The Big Idea** How was the distance-time graph useful for describing the motion of the ball?

## Communicate Your Results

Prepare a poster that shows your graph and describes how it can be used to calculate average speed.

Design and conduct an experiment comparing the average speed of different types of balls along the track.

**8**

| Distance (m) | Time(s) | | |
|---|---|---|---|
| | 2 books | 3 books | 4 books |
| 0.50 | | | |
| 1.00 | | | |
| 1.50 | | | |
| 2.00 | | | |
| 2.50 | | | |
| 3.00 | | | |

## Lab Tips

☑ If the ball doesn't roll far enough, reduce the track length to 2 m.

☑ Practice using the stopwatches several times to gain experience in making accurate readings.

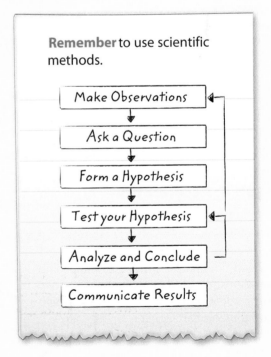

**Remember** to use scientific methods.

Make Observations

Ask a Question

Form a Hypothesis

Test your Hypothesis

Analyze and Conclude

Communicate Results

# Chapter 1 Study Guide

**THE BIG IDEA**

The motion of an object can be described by the object's position, velocity, and acceleration.

| Key Concepts Summary 🗝 | Vocabulary |
|---|---|
| **Lesson 1: Position and Motion**<br><br>• An object's **position** is its distance and direction from a **reference point.**<br><br>• The position of an object in two dimensions can be described by choosing a reference point and two reference directions, and then stating the distance along each reference direction.<br><br>• The distance an object moves is the actual length of its path. Its **displacement** is the difference between its initial position and its final position.<br><br> | **reference point** p. 9<br>**position** p. 9<br>**motion** p. 13<br>**displacement** p. 13 |
| **Lesson 2: Speed and Velocity**<br><br>• **Speed** is the distance an object moves per unit of time.<br><br>• An object moving the same distance each second is moving at a **constant speed.** The speed of an object at a certain moment is its **instantaneous speed.** <br><br>• You can calculate an object's **average speed** from a distance-time graph by dividing the distance the object travels by the total time it takes to travel that distance.<br><br>• **Velocity** changes when speed, direction, or both speed and direction change. | **speed** p. 17<br>**constant speed** p. 18<br>**instantaneous speed** p. 18<br>**average speed** p. 19<br>**velocity** p. 23 |
| **Lesson 3: Acceleration**<br><br>• **Acceleration** is a change in velocity over time. An object accelerates when it speeds up, slows down, or changes direction.<br><br>• A speed-time graph shows the relationship between speed and time and can be used to determine information about the acceleration of an object.<br><br> | **acceleration** p. 27 |

## FOLDABLES® Chapter Project

Assemble your lesson Foldables as shown to make a Chapter Project. Use the project to review what you have learned in this chapter.

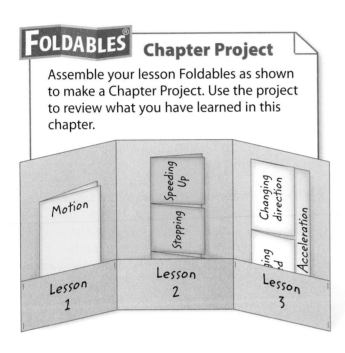

## Use Vocabulary

**①** A pencil's _____ might be described as 3 cm to the left of the stapler.

**②** An object that changes position is in _____.

**③** If an object is traveling at a _____, it does not speed up or slow down.

**④** An object's _____ includes both its speed and the direction it moves.

**⑤** An object's change in velocity during a time interval, divided by the time interval during which the velocity changed, is its _____.

**⑥** A truck driver stepped on the brakes to make a quick stop. The truck's _____ is in the opposite direction as its velocity.

## Link Vocabulary and Key Concepts

 **((○)) Concepts in Motion**   Interactive Concept Map

*Copy this concept map, and then use vocabulary terms from the previous page to complete the concept map.*

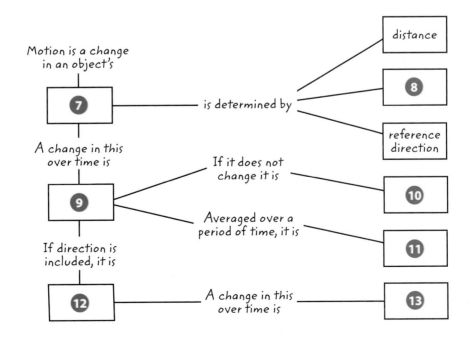

## Understand Key Concepts

**1** An airplane rolls down the runway. Compared to which reference point is the airplane in motion?
   A. the cargo the plane carries
   B. the control tower
   C. the pilot flying the plane
   D. the plane's wing

**2** Which describes motion in two dimensions?
   A. a car driving through a city
   B. a rock dropping off a cliff
   C. a sprinter on a 100-m track
   D. a train on a straight track

**3** Which line represents the greatest average speed during the 30-s time period?

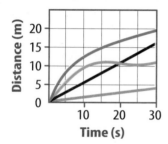

   A. the blue line
   B. the black line
   C. the green line
   D. the orange line

**4** Which describes the greatest displacement?
   A. walking 3 m east, then 3 m north, then 3 m west
   B. walking 3 m east, then 3 m south, then 3 m east
   C. walking 3 m north, then 3 m south, then 3 m north
   D. walking 3 m north, then 3 m west, then 3 m south

**5** Which has the greatest average speed?
   A. a boat sailing 80 km in 2 hours
   B. a car driving 90 km in 3 hours
   C. a train traveling 120 km in 3 hours
   D. a truck moving 50 km in 1 hour

**6** Which describes motion in which the person or object is accelerating?
   A. A bird flies straight from a tree to the ground without changing speed.
   B. A dog walks at a constant speed along a straight sidewalk.
   C. A girl runs along a straight path the same distance each second.
   D. A truck moves around a curve without changing speed.

**7** Richard walks from his home to his school at a constant speed. It takes him 4 min to travel 100 m. Which of the lines in the following distance-time graph could show Richard's motion on the way to school?

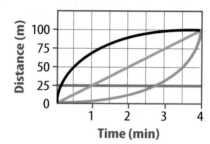

   A. the black line
   B. the blue line
   C. the green line
   D. the orange line

**8** Which is a unit of acceleration?
   A. $kg/m$
   B. $kg \cdot m/s^2$
   C. $m/s$
   D. $m/s^2$

**9** Which have the same velocity?
   A. a boy walking east at 2 km/h and a man walking east at 4 km/h
   B. a car standing still and a truck driving in a circle at 4 km/h
   C. a dog walking west at 3 km/h and a cat walking west at 3 km/h
   D. a girl walking west at 3 km/h and a boy walking south at 3 km/h

## Critical Thinking

**10** **Describe** A ruler is on the table with the higher numbers to the right. An ant crawls along the ruler from 6 cm to 2 cm in 2 seconds. Describe the ant's distance, displacement, speed, and velocity.

**11** **Describe** a theme-park ride that has constant speed but changing velocity.

**12** **Construct** a distance-time graph that shows the following motion: A person leaves a starting point at a constant speed of 4 m/s and walks for 4 s. The person then stops for 2 s. The person then continues walking at a constant speed of 2 m/s for 4 s.

**13** **Calculate** A truck driver travels 55 km in 1 hour. He then drives a speed of 35 km/h for 2 hours. Next, he drives 175 km in 3 hours. What was his average speed?

**14** **Interpret** Keisha measured the distance her friend Morgan ran on a straight track every 2 s. Her measurements are recorded in the table below. What was Morgan's average speed? What was her acceleration?

| Time (s) | Distance (m) |
|----------|--------------|
| 0        | 0            |
| 2        | 2            |
| 4        | 6            |
| 6        | 8            |
| 8        | 14           |
| 10       | 20           |

*Writing in Science*

**15** **Write** A friend tells you he is 30 m from the fountain in the middle of the city. Write a short paragraph explaining why you cannot identify your friend's position from this description.

**REVIEW**  **THE BIG IDEA**

**16** Nora rides a bicycle for 5 min on a curvy road at a constant speed of 10 m/s. Describe Nora's ride in terms of position, velocity, and acceleration. Compare the distance she rides and her displacement.

**17** What are some ways to describe the motion of the jets in the photograph below?

**Math Skills**

**Review**
Math Practice

### Solve One-Step Equations

**18** A model train moves 18.3 m in 122 s. What is the train's average speed?

**19** A car travels 45 km in an hour. In each of the next two hours, it travels 78 km. What is the average speed of the car?

**20** The speed of a car traveling on a straight road increases from 63 m/s to 75 m/s in 4.2 s. What is the car's acceleration?

**21** A girl starts from rest and reaches a walking speed of 1.4 m/s in 3.0 s. She walks at this speed for 6.0 s. The girl then slows down and comes to a stop during a 10.0-s period. What was the girl's acceleration during each of the three time periods? What was her acceleration for the entire trip?

# Standardized Test Practice

*Record your answers on the answer sheet provided by your teacher or on a sheet of paper.*

## Multiple Choice

**1** Radar tells an air traffic controller that a jet is slowing as it nears the airport. Which might represent the jet's speed?

   **A** 700 h

   **B** 700 h/km

   **C** 700 km

   **D** 700 km/h

*Use the diagram below to answer question 2.*

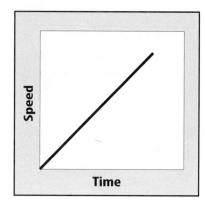

**2** What does the graph above illustrate?

   **A** average speed

   **B** constant speed

   **C** decreasing speed

   **D** increasing speed

**3** Why is a car accelerating when it is circling at a constant speed?

   **A** It is changing its destination.

   **B** It is changing its direction.

   **C** It is changing its distance.

   **D** It is changing its total mass.

**4** Which is defined as the process of changing position?

   **A** displacement

   **B** distance

   **C** motion

   **D** relativity

**5** Each diagram below shows two sliding boxes. Which boxes have the same velocity?

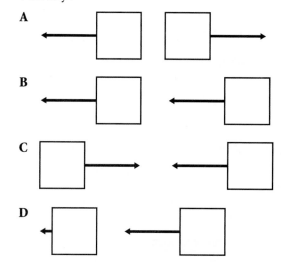

**6** In the phrase "two miles southeast of the mall," what is the mall?

   **A** a dimension

   **B** a final destination

   **C** a position

   **D** a reference point

**7** The initial speed of a dropped ball is 0 m/s. After 2 seconds, the ball travels at a speed of 20 m/s. What is the acceleration of the ball?

   **A** 5 m/s$^2$

   **B** 10 m/s$^2$

   **C** 20 m/s$^2$

   **D** 40 m/s$^2$

**8** Which could be described by the expression "100 m/s northwest"?

 **A** acceleration

 **B** distance

 **C** speed

 **D** velocity

*Use the diagram below to answer question 9.*

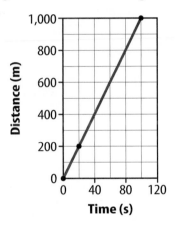

**9** In the above graph, what is the average speed of the moving object between 20 and 60 seconds?

 **A** 5 m/s

 **B** 10 m/s

 **C** 20 m/s

 **D** 40 m/s

**10** A car travels 250 km and stops twice along the way. The entire trip takes 5 hours. What is the average speed of the car?

 **A** 25 km/h

 **B** 40 km/h

 **C** 50 km/h

 **D** 250 km/h

## Constructed Response

*Use the diagram below to answer questions 11 and 12.*

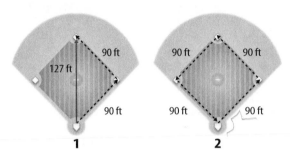

**11** The dashed lines show the paths two players run on baseball diamonds. What distance does the player travel on diamond 1? How does it compare to the distance the player runs on diamond 2?

**12** Calculate the displacement of the runners on diamonds 1 and 2. Explain your answers.

*Use the diagram below to answer question 13.*

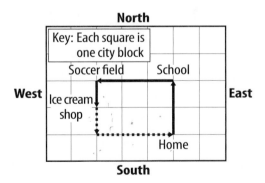

**13** A student walks from home to school, on to a soccer field, then to an ice cream shop, and finally home. Use grid distances and directions to describe each leg of his trip. What is the distance between the student's home and the ice cream shop?

| NEED EXTRA HELP? | | | | | | | | | | | | | |
|---|---|---|---|---|---|---|---|---|---|---|---|---|---|
| If You Missed Question... | 1 | 2 | 3 | 4 | 5 | 6 | 7 | 8 | 9 | 10 | 11 | 12 | 13 |
| Go to Lesson... | 2 | 3 | 3 | 1 | 2 | 1 | 3 | 2 | 2 | 2 | 1 | 1 | 1 |

# Chapter 2

# The Laws of Motion

**THE BIG IDEA**

How do forces change the motion of objects?

**Inquiry** **Why move around?**

Imagine the sensations these riders experience as they swing around. The force of gravity pulls the riders downward. Instead of falling, however, they move around in circles.

- What causes the riders to move around?
- What prevents the riders from falling?
- How do forces change the motion of the riders?

## Get Ready to Read

### What do you think?

Before you read, decide if you agree or disagree with each of these statements. As you read this chapter, see if you change your mind about any of the statements.

1 You pull on objects around you with the force of gravity.

2 Friction can act between two unmoving, touching surfaces.

3 Forces acting on an object cannot be added.

4 A moving object will stop if no forces act on it.

5 When an object's speed increases, the object accelerates.

6 If an object's mass increases, its acceleration also increases if the net force acting on the object stays the same.

7 If objects collide, the object with more mass applies more force.

8 Momentum is a measure of how hard it is to stop a moving object.

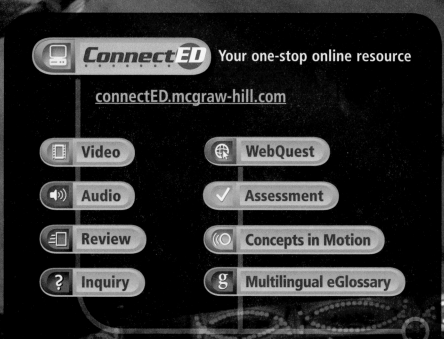

**ConnectED**  Your one-stop online resource

connectED.mcgraw-hill.com

| Video | WebQuest |
| Audio | Assessment |
| Review | Concepts in Motion |
| Inquiry | Multilingual eGlossary |

# Lesson 1

## Reading Guide

**Key Concepts**
**ESSENTIAL QUESTIONS**

- What are some contact forces and some noncontact forces?

- What is the law of universal gravitation?

- How does friction affect the motion of two objects sliding past each other?

**Vocabulary**

**force** p. 45

**contact force** p. 45

**noncontact force** p. 46

**gravity** p. 47

**mass** p. 47

**weight** p. 48

**friction** p. 49

**g** **Multilingual eGlossary**

**Video**

**What's Science Got to do With It?**

# Gravity and Friction

## Inquiry Why doesn't he fall?

This astronaut is on an aircraft that flies at steep angles and provides a sense of weightlessness. Why doesn't he fall? He does! Earth's gravity pulls the astronaut down, but the aircraft moves downward at the same speed.

## Can you make a ball move without touching it?

You can make a ball move by kicking it or throwing it. Is it possible to make the ball move even when nothing is touching the ball?

1. Read and complete a lab safety form.

2. Roll a **tennis ball** across the floor. Think about what makes the ball move.

3. Toss the ball into the air. Watch as it moves up and then falls back to your hand.

4. Drop the ball onto the floor. Let it bounce once, and then catch it.

**Think About This**

1. What made the ball move when you rolled, tossed, and dropped it? What made it stop?

2. 🔑 **Key Concept** Did something that was touching the ball or not touching the ball cause it to move in each case?

## Types of Forces

Think about all the things you pushed or pulled today. You might have pushed toothpaste out of a tube. Maybe you pulled out a chair to sit down. *A push or a pull on an object is called a* **force.** An object or a person can apply a force to another object or person. Some forces are applied only when objects touch. Other forces are applied even when objects do not touch.

### Contact Forces

The hand of the karate expert in **Figure 1** applied a force to the stack of boards and broke them. You have probably also seen a musician strike the keys of a piano and an athlete hit a ball with a bat. In each case, a person or an object applied a force to an object that it touched. *A* **contact force** *is a push or a pull on one object by another that is touching it.*

Contact forces can be weak, like when you press the keys on a computer keyboard. They also can be strong, such as when large sections of underground rock suddenly move, resulting in an earthquake. The large sections of Earth's crust called plates also apply strong contact forces against each other. Over long periods of time, these forces can create mountain ranges if one plate pushes another plate upward.

**WORD ORIGIN** ············

force
from Latin *fortis*, means "strong"

**Figure 1** The man's hand applies a contact force to the boards.

▲ **Figure 2** A noncontact force causes the girl's hair to stand on end.

## Noncontact Forces

Lift a pencil and then release it. What happens? The pencil falls toward the floor. A parachutist falls toward Earth even though nothing is touching him. *A force that one object can apply to another object without touching it is a* **noncontact force.** Gravity, which pulled on your pencil and the parachutist, is a noncontact force. The magnetic force, which attracts certain metals to magnets, is also a noncontact force. In **Figure 2,** another noncontact force, called the electric force, causes the girl's hair to stand on end.

**Key Concept Check** What are some contact forces and some noncontact forces?

## Strength and Direction of Forces

Forces have both strength and direction. If you push your textbook away from you, it probably slides across your desk. What happens if you push down on your book? It probably does not move. You can use the same strength of force in both cases. Different things happen because the direction of the applied force is different.

As shown in **Figure 3,** arrows can be used to show forces. The length of an arrow shows the strength of the force. Notice in the figure that the force applied by the tennis racquet is stronger than the force applied by the table-tennis paddle. As a result, the arrow showing the force of the tennis racquet is longer. The direction that an arrow points shows the direction in which force was applied.

The SI unit for force is the newton (N). You apply a force of about 1 N when lifting a stick of butter. You use a force of about 20 N when lifting a 2-L bottle of water. If you use arrows to show these forces, the water's arrow would be 20 times longer.

**Figure 3** Arrows can indicate the strength and direction of a force. ▼

300 N

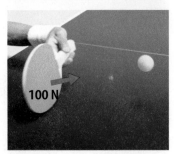

100 N

**Visual Check** How are the lengths of the arrows related to the different forces on the two balls?

| Change in Mass | Change in Distance |
|---|---|
| 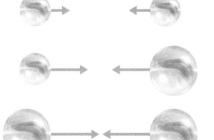 | |
| Gravitational force increases if the mass of at least one of the objects increases. | The gravitational force between objects decreases as the objects move apart. |

## What is gravity?

Objects fall to the ground because Earth exerts an attractive force on them. Did you know that you also exert an attractive force on objects? **Gravity** *is an attractive force that exists between all objects that have mass.* **Mass** *is the amount of matter in an object.* Mass is often measured in kilograms (kg).

### The Law of Universal Gravitation

In the late 1600s, an English scientist and mathematician, Sir Isaac Newton, developed the law of universal gravitation. This law states that all objects are attracted to each other by a gravitational force. The strength of the force depends on the mass of each object and the distance between them.

 **Key Concept Check** What is the law of universal gravitation?

**Gravitational Force and Mass** The way in which the mass of objects affects gravity is shown in **Figure 4.** When the mass of one or both objects increases, the gravitational force between them also increases. Notice that the force arrows for each pair of marbles are the same size even when one object has less mass. Each object exerts the same attraction on the other object.

**Gravitational Force and Distance** The effect that distance has on gravity is also shown in **Figure 4.** The attraction between objects decreases as the distance between the objects increases. For example, if your mass is 45 kg, the gravitational force between you and Earth is about 440 N. On the Moon, about 384,000 km away, the gravitational force between you and Earth would only be about 0.12 N. The relationship between gravitational force and distance is shown in the graph in **Figure 5.**

 **Reading Check** What effect does distance have on gravity?

▲ **Figure 4**  The gravitational force between objects depends on the mass of the objects and the distance between them.

**Review**
**Personal Tutor**

**Effect of Distance on Gravity**

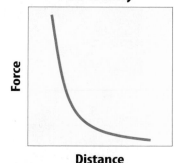

▲ **Figure 5** The gravitational force between objects decreases as the distance between the objects increases.

## Weight—A Gravitational Force

Earth has more mass than any object near you. As a result, the gravitational force Earth exerts on you is greater than the force exerted by any other object. **Weight** *is the gravitational force exerted on an object.* Near Earth's surface, an object's weight is the gravitational force exerted on the object by Earth. Because weight is a force, it is measured in newtons.

**The Relationship Between Weight and Mass** An object's weight is proportional to its mass. For example, if one object has twice the mass of another object, it also has twice the weight. Near Earth's surface, the weight of an object in newtons is about ten times its mass in kilograms.

 **Reading Check** What is the relationship between mass and weight?

**ACADEMIC VOCABULARY**

**significant**
(***adjective***) important, momentous

**Weight and Mass High Above Earth** You might think that astronauts in orbit around Earth are weightless. Their weight is about 90 percent of what it is on Earth. The mass of the astronaut in **Figure 6** is about 55 kg. Her weight is about 540 N on Earth and about 500 N on the space station 350 km above Earth's surface. Why is there no **significant** change in weight when the distance increases so much? Earth is so large that an astronaut must be much farther away for the gravitational force to change much. The distance between the astronaut and Earth is small compared to the size of Earth.

 **Reading Check** Why is the gravitational force that a friend exerts on you less than the gravitational force exerted on you by Earth?

**Figure 6** As she travels from Earth to the space station, the astronaut's weight changes, but her mass remains the same.

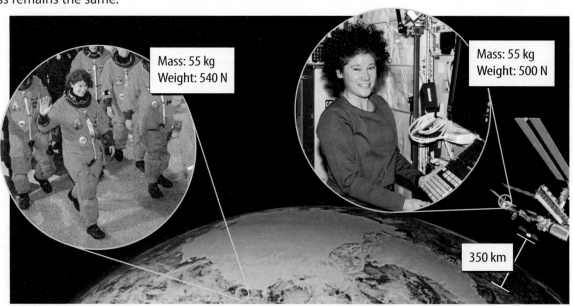

Mass: 55 kg
Weight: 540 N

Mass: 55 kg
Weight: 500 N

350 km

**Visual Check** What would be the weight of a 110-kg object on Earth? On the space station?

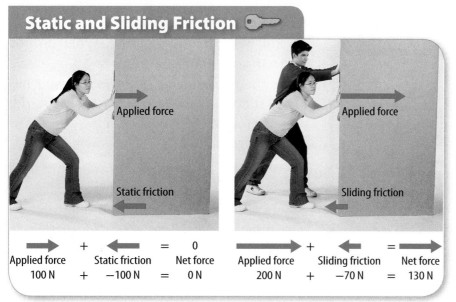

## Static and Sliding Friction 🔑

Applied force

Static friction

Applied force

Sliding friction

| Applied force | + | Static friction | = | Net force |
|---|---|---|---|---|
| 100 N | + | −100 N | = | 0 N |

| Applied force | + | Sliding friction | = | Net force |
|---|---|---|---|---|
| 200 N | + | −70 N | = | 130 N |

◄ **Figure 7** Static friction prevents the box on the left from moving. Sliding friction slows the motion of the box on the right.

# Friction

If you slide across a smooth floor in your socks, you move quickly at first and then stop. The force that slows you is friction. **Friction** *is a force that resists the motion of two surface that are touching.* There are several types of friction.

## Static Friction

The box on the left in **Figure 7** does not move because the girl's applied force is balanced by **static** friction. Static friction prevents surfaces from sliding past each other. Up to a limit, the strength of static friction changes to match the applied force. If the girl increases the applied force, the box still will not move because the static friction also increases.

## Sliding Friction

When static friction reaches its limit between surfaces, the box will move. As shown in **Figure 7,** the force of two students pushing is greater than the static friction between the box and the floor. Sliding friction opposes the motion of surfaces sliding past each other. As long as the box is sliding, the sliding friction does not change. Increasing the applied force makes the box slide faster. If the students stop pushing, the box will slow and stop because of sliding friction.

## Fluid Friction

Friction between a surface and a fluid—any material, such as water or air, that flows—is fluid friction. Fluid friction between a surface and air is air resistance. Suppose an object is moving through a fluid. Decreasing the surface area toward the oncoming fluid decreases the air resistance against the object. The crumpled paper in **Figure 8** falls faster than the flat paper because it has less surface area and less air resistance.

SCIENCE USE V. COMMON USE · ·
**static**
*Science Use* at rest or having no motion

*Common Use* noise produced in a radio or a television

**Figure 8** Air resistance is greater on the flat paper. ▼

Air resistance

Gravity

Air resistance

Gravity

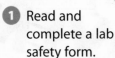
### How does friction affect motion?

Friction affects the motion of an object sliding across a surface.

1 Read and complete a lab safety form.

2 Use **tape** to fasten **sandpaper** to a table. Attach a **spring scale** to a **wooden block** with an **eyehook** in it.

3 Record in your Science Journal the force required to gently pull the block at a constant speed on the sandpaper and then on the table.

**Analyze and Conclude**

1. **Compare** the forces required to pull the block across the two surfaces.

2.  **Key Concept** How did reducing friction affect the motion of the block?

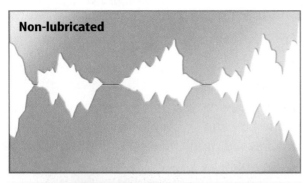

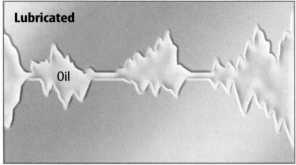

**Figure 9** Lubricants such as oil decrease friction caused by microscopic roughness.

## What causes friction?

Rub your hands together. What do you feel? If your hands were soapy, you could slide them past each other easily. You feel more friction when you rub your dry hands together than when you rub your soapy hands together.

What causes friction between surfaces? Look at the close-up view of surfaces in **Figure 9.** Microscopic dips and bumps cover all surfaces. When surfaces slide past each other, the dips and bumps on one surface catch on the dips and bumps on the other surface. This microscopic roughness slows sliding and is a source of friction.

**Key Concept Check** How does friction affect the motion of two objects sliding past each other?

In addition, small particles—atoms and molecules—make up all surfaces. These particles contain weak electrical charges. When a positive charge on one surface slides by a negative charge on the other surface, an attraction occurs between the particles. This attraction slows sliding and is another source of friction between the surfaces.

**Reading Check** What are two causes of friction?

## Reducing Friction

When you rub soapy hands together, the soapy water slightly separates the surfaces of your hands. There is less contact between the microscopic dips and bumps and between the electrical charges of your hands. Soap acts as a lubricant and decreases friction. With less friction, it is easier for surfaces to slide past each other, as shown in **Figure 9.** Motor oil is a lubricant that reduces friction between moving parts of a car's engine.

Look again at the effect of air resistance on the falling paper in **Figure 8.** Reducing the paper's surface area reduces the fluid friction between it and the air.

## Visual Summary

Forces can be either contact, such as a karate chop, or noncontact, such as gravity. Each type is described by its strength and direction.

Gravity is an attractive force that acts between any two objects that have mass. The attraction is stronger for objects with greater mass.

Friction can reduce the speed of objects sliding past each other. Air resistance is a type of fluid friction that slows the speed of a falling object.

**FOLDABLES**

Use your lesson Foldable to review the lesson. Save your Foldable for the project at the end of the chapter.

## What do you think NOW?

You first read the statements below at the beginning of the chapter.

**1.** You pull on objects around you with the force of gravity.

**2.** Friction can act between two unmoving, touching surfaces.

Did you change your mind about whether you agree or disagree with the statements? Rewrite any false statements to make them true.

## Use Vocabulary

**1** **Define** *friction* in your own words.

**2** **Distinguish** between weight and mass.

## Understand Key Concepts

**3** **Explain** the difference between a contact force and a noncontact force.

**4** You push a book sitting on a desk with a force of 5 N, but the book does not move. What is the static friction?
  **A.** 0 N              **C.** between 0 N and 5 N
  **B.** 5 N              **D.** greater than 5 N

**5** **Apply** According to the law of universal gravitation, is there a stronger gravitational force between you and Earth or an elephant and Earth? Why?

## Interpret Graphics

**6** **Interpret** Look at the forces on the feather.

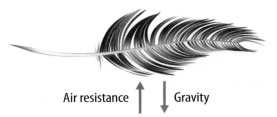

Air resistance ↑  ↓ Gravity

In terms of these forces, explain why the feather falls slowly rather than fast.

**7** **Organize Information** Copy and fill in the table below to describe forces mentioned in this lesson. Add as many rows as you need.

| Force | Description |
|-------|-------------|
|       |             |
|       |             |
|       |             |

## Critical Thinking

**8** **Decide** Is it possible for the gravitational force between two 50-kg objects to be less than the gravitational force between a 50-kg object and a 5-kg object? Explain.

AMERICAN MUSEUM ℠ NATURAL HISTORY

# Avoiding an Asteroid Collision

The force of gravity can change the path of an asteroid moving through the solar system.

## Gravity to the rescue!

Everything in the universe—from asteroids to planets to stars—exerts gravity on every other object. This force keeps the Moon in orbit around Earth and Earth in orbit around the Sun. Gravity can also send objects on a collision course—a problem when those objects are Earth and an asteroid. Asteroids are rocky bodies found mostly in the asteroid belt between Mars and Jupiter. Jupiter's strong gravity can change the orbits of asteroids over time, occasionally sending them dangerously close to Earth.

Astronomers use powerful telescopes to track asteroids near Earth. More than a thousand asteroids are large enough to cause serious damage if they collide with Earth. If an asteroid were heading our way, how could we prevent the collision? One idea is to launch a spacecraft into the asteroid. The impact could slow it down enough to cause it to miss Earth. But if the asteroid broke apart, the pieces could rain down onto Earth!

Scientists have another idea. They propose launching a massive spacecraft into an orbit close to the asteroid. The spacecraft's gravity would exert a small tug on the asteroid. Over time, the asteroid's path would be altered enough to pass by Earth. Astronomers track objects now that are many years away from crossing paths with Earth. This gives them enough time to set a plan in motion if one of the objects appears to be on a collision course with Earth.

The Spacewatch telescope in Arizona scans the sky for near-Earth asteroids. Other U.S. telescopes with this mission are in Hawaii, California, and New Mexico.

Meteor Crater in Arizona was created when an asteroid about 50 m wide collided with Earth about 50,000 years ago.

## It's Your Turn

**PROBLEM SOLVING** With a group, come up with a plan for avoiding an asteroid's collision with Earth. Present your plan to the class. Include diagrams and details that explain exactly how your plan will work.

# Lesson 2

## Reading Guide

### Key Concepts
**ESSENTIAL QUESTIONS**

- What is Newton's first law of motion?
- How is motion related to balanced and unbalanced forces?
- What effect does inertia have on the motion of an object?

### Vocabulary
**net force** p. 55

**balanced forces** p. 56

**unbalanced forces** p. 56

**Newton's first law of motion** p. 57

**inertia** p. 58

 **g** Multilingual eGlossary

**Video** BrainPOP®

# Newton's First Law

## inquiry How does it balance?

You probably would be uneasy standing under Balanced Rock near Buhl, Idaho. Yet this unusual rock stays in place year after year. The rock has forces acting on it. Why doesn't it fall over? The forces acting on the rock combine, and the rock does not move.

## Can you balance magnetic forces?

Magnets exert forces on each other. Depending on how you hold them, magnets either attract or repel each other. Can you balance these magnetic forces?

1. Read and complete a lab safety form.

2. Have your lab partner hold a **ring magnet** vertically on a **pencil,** as shown in the picture.

3. Place **another magnet** on the pencil, and use it to push the first magnet along the pencil.

4. Place a **third magnet** on the same pencil so that the outer magnets push against the middle one. Does the middle magnet still move along the pencil?

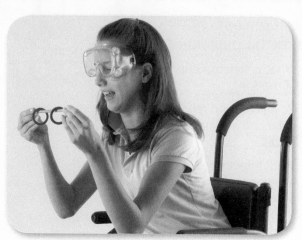

### Think About This

1. Describe the forces that the other magnets exert on the first magnet in steps 3 and 4.

2. 🔑 **Key Concept** Describe how the motion of the first magnet seemed to depend on whether each force on the magnet was balanced by another force.

## Identifying Forces

Ospreys are birds of prey that live near bodies of water. Perhaps several minutes ago, the mother osprey in **Figure 10** was in the air in a high-speed dive. It might have plunged toward a nearby lake after seeing a fish in the water. As it neared the water, it moved its legs forward to grab the fish with its talons. It then stretched out its wings and used them to climb high into the air. Before the osprey comes to rest on its nest, it will slow its speed and land softly on the nest's edge, near the young birds waiting for food.

Forces helped the mother osprey change the speed and direction of its motion. Recall that a force is a push or a pull. Some of the forces were contact forces, such as air resistance. When soaring, the osprey spread its wings, increasing air resistance. In a dive, it held its wings close to its body, changing its shape, decreasing its surface area and air resistance. Gravity also pulled the osprey toward the ground.

To understand the motion of an object, you need to identify the forces acting on it. In this lesson you will read how forces change the motion of objects.

**Figure 10** Forces change the motion of this osprey.

## Combining Forces—The Net Force

Suppose you try to move a piece of heavy furniture, such as the dresser in **Figure 11.** If you push on the dresser by yourself, you have to push hard on the dresser to overcome the static friction and move it. If you ask a friend to push with you, you do not have to push as hard. When two or more forces act on an object, the forces combine. *The combination of all the forces acting on an object is the* **net force.** The way in which forces combine depends on the directions of the forces applied to an object.

## Combining Forces in the Same Direction

When the forces applied to an object act in the same direction, the net force is the sum of the individual forces. In this case, the direction of the net force is the same as the direction of the individual forces.

Because forces have direction, you have to specify a **reference direction** when you combine forces. In **Figure 11,** for example, you would probably choose "to the right" as the positive reference direction. Both forces then would be positive. The net force on the dresser is the sum of the two forces pushing in the same direction. One person pushes on the dresser with a force of 200 N to the right. The other person pushes with a force of 100 N to the right. The net force on the dresser is 200 N + 100 N = 300 N to the right. The force applied to the dresser is the same as if one person pushed on the dresser with a force of 300 N to the right.

**Reading Check** How do you calculate the net force on an object if two forces are acting on it in the same direction?

**REVIEW VOCABULARY**

**reference direction**
a direction that you choose from a starting point to describe an object's position

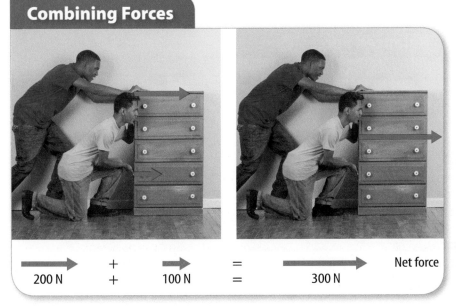

### Combining Forces

| →  200 N | + | →  100 N | = | →  300 N | Net force |

**Figure 11** When forces in the same direction combine, the net force is also in the same direction. The strength of the net force is the sum of the forces.

**Visual Check** What would the net force be if one boy pushed with 250 N and the other boy pushed in the same direction with 180 N?

Review    Personal Tutor

**Figure 12** When two forces acting on an object in opposite directions combine, the net force is in the same direction as the larger force. The strength of the net force is the sum of the positive and negative forces. ▶

**Unbalanced Forces** 🔑

| | | | | | |
|---|---|---|---|---|---|
| 200 N | + | −100 N | = | 100 N | Net force |

## Combining Forces in Opposite Directions

When forces act in opposite directions on an object, the net force is still the sum of the forces. Suppose you choose "to the right" again as the reference direction in **Figure 12.** A force in that direction is positive, and a force in the opposite direction is negative. The net force is the sum of the positive and negative forces. The net force on the dresser is 100 N to the right.

## Balanced and Unbalanced Forces

When equal forces act on an object in opposite directions, as in **Figure 13,** the net force on the object is zero. The effect is the same as if there were no forces acting on the object. *Forces acting on an object that combine and form a net force of zero are* **balanced forces.** Balanced forces do not change the motion of an object. However, the net force on the dresser in **Figure 12** is not zero. There is a net force to the right. *Forces acting on an object that combine and form a net force that is not zero are* **unbalanced forces.**

**Figure 13** When two forces acting on an object in opposite directions are the same strength, the forces are balanced. ▶

✅ **Visual Check** How are the force arrows for the balanced forces in the figure alike? How are they different?

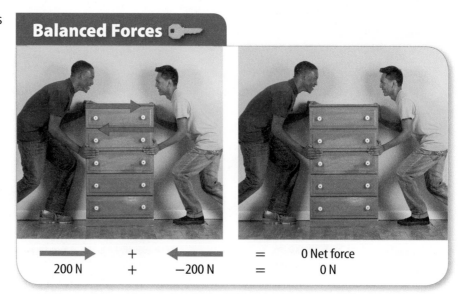

**Balanced Forces** 🔑

| | | | | |
|---|---|---|---|---|
| 200 N | + | −200 N | = | 0 Net force |
| | | | | 0 N |

**56** Chapter 2
EXPLAIN

## Newton's First Law of Motion

Sir Isaac Newton studied how forces affect the motion of objects. He developed three rules known as Newton's laws of motion. *According to* **Newton's first law of motion,** *if the net force on an object is zero, the motion of the object does not change.* As a result, balanced forces and unbalanced forces have different results when they act on an object.

 **Key Concept Check** What is Newton's first law of motion?

## Balanced Forces and Motion

According to Newton's first law of motion, balanced forces cause no change in an object's velocity (speed in a certain direction). This is true when an object is at rest or in motion. Look again at **Figure 13.** The dresser is at rest before the boys push on it. It remains at rest when they apply balanced forces. Similarly, because the forces in **Figure 14**—air resistance and gravity—are balanced, the parachutist moves downward at his terminal velocity. Terminal velocity is the constant velocity reached when air resistance equals the force of gravity acting on a falling object.

 **Reading Check** What happens to the velocity of a moving car if the forces on it are balanced?

## Unbalanced Forces and Motion

Newton's first law of motion only applies to balanced forces acting on an object. When unbalanced forces act on an object at rest, the object starts moving. When unbalanced forces act on an already moving object, the object's speed, direction of motion, or both change. You will read more about how unbalanced forces affect an object's motion in the next lesson.

 **Key Concept Check** How is motion related to balanced and unbalanced forces?

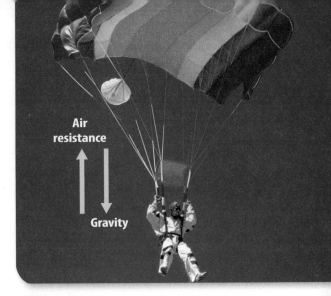

**Air resistance**

**Gravity**

**Figure 14** Balanced forces acting on an object do not change the object's speed and direction.

**Inquiry** **MiniLab** **15 minutes**

### How do forces affect motion?

The motion of an object depends on whether balanced or unbalanced forces act on it.

1. Read and complete a lab safety form.
2. Attach **spring scales** to opposite sides of a **wooden block** with **eyehooks.**
3. With a partner, gently pull the scales so that the block moves toward one of you. Sketch the setup in your Science Journal, including the force readings on each scale.
4. Repeat step 3, pulling on the block so that it does not move.

**Analyze and Conclude**

1. **Explain** Use Newton's first law of motion to explain what occurred in steps 3 and 4.
2.  **Key Concept** How was the block's motion related to balanced and unbalanced forces?

✅ **Visual Check** What effect would a shoulder belt and a lap belt have on the inertia of the crash-test dummy?

### Inertia

According to Newton's first law, the motion of an object will not change if balanced forces act on it. *The tendency of an object to resist a change in its motion is called* **inertia** (ihn UR shuh). Inertia explains the motion of the crash-test dummy in **Figure 15.** Before the crash, the car and dummy moved with constant velocity. If no other force had acted on them, the car and dummy would have continued moving with constant velocity because of inertia. The impact with the barrier results in an unbalanced force on the car, and the car stops. The dummy continues moving forward because of its inertia.

🔑 **Key Concept Check** What effect does inertia have on the motion of an object?

## Why do objects stop moving?

Think about how friction and inertia together affect an object's movement. A book sitting on a table, for example, stays in place because of inertia. When you push the book, the force you apply to the book is greater than static friction between the book and the table. The book moves in the direction of the greater force. If you stop pushing, friction stops the book.

What would happen if there were no friction between the book and the table? Inertia would keep the book moving. According to Newton's first law, the book would continue to move at the same speed in the same direction as your push.

On Earth, friction can be reduced but not totally removed. For an object to start moving, a force greater than static friction must be applied to it. To keep the object in motion, a force at least as strong as friction must be applied continuously. Objects stop moving because friction or another force acts on them.

**WORD ORIGIN** ·············

**inertia**
from Latin *iners*, means "without skill, inactive"

**FOLDABLES**

Make a chart with six columns and six rows. Use your chart to define and show how this lesson's vocabulary terms are related. Afterward, fold your chart in half and label the outside *Newton's First Law.*

| | Net Force | Balanced Forces | Unbalanced Forces | Newton's First Law | Inertia |
|---|---|---|---|---|---|
| Net Force | | | | | |
| Balanced Forces | | | | | |
| Unbalanced Forces | | | | | |
| Newton's First Law | | | | | |
| Inertia | | | | | |

# Lesson 2 Review

## Visual Summary

Unbalanced forces cause an object to move.

According to Newton's first law of motion, if the net force on an object is zero, the object's motion does not change.

Inertia is a property that resists a change in the motion of an object.

**FOLDABLES®**

Use your lesson Foldable to review the lesson. Save your Foldable for the project at the end of the chapter.

## What do you think NOW?

You first read the statements below at the beginning of the chapter.

**3.** Forces acting on an object cannot be added.

**4.** A moving object will stop if no forces act on it.

Did you change your mind about whether you agree or disagree with the statements? Rewrite any false statements to make them true.

## Use Vocabulary

**1** **Define** *net force* in your own words.

**2** **Distinguish** between balanced forces and unbalanced forces.

## Understand Key Concepts

**3** Which causes an object in motion to remain in motion?
-  **A.** friction
-  **C.** inertia
-  **B.** gravity
-  **D.** velocity

**4** **Apply** You push a coin across a table. The coin stops. How does this motion relate to balanced and unbalanced forces?

**5** **Explain** Use Newton's first law to explain why a book on a desk does not move.

## Interpret Graphics

**6** **Analyze** What is the missing force?

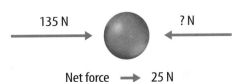

135 N          ? N

Net force ⟶ 25 N

**7** **Organize Information** Copy and fill in the graphic organizer below to explain Newton's first law of motion in each case.

| Object at rest | |
|---|---|
| Object in motion | |

## Critical Thinking

**8** **Extend** Three people push a piano on wheels with forces of 130 N to the right, 150 N to the left, and 165 N to the right. What are the strength and direction of the net force on the piano?

**9** **Assess** A child pushes down on a box lid with a force of 25 N. At the same time, her friend pushes down on the lid with a force of 30 N. The spring on the box lid pushes upward with a force of 60 N. Can the children close the box? Why or why not?

# How can you model Newton's first law of motion?

**Materials**

markers
(red, blue,
black, green)

According to Newton's first law of motion, balanced forces do not change an object's motion. Unbalanced forces change the motion of objects at rest or in motion. You can model different forces and their effects on the motion of an object.

## Learn It

When you **model** a concept in science, you act it out, or imitate it. You can model the effect of balanced and unbalanced forces on motion by using movements on a line.

## Try It

1. Draw a line across a sheet of lined notebook paper lengthwise. Place an X at the center. Each space to the right of the X will model a force of 1 N east, and each space to the left will model 1 N west.

2. Suppose a force of 3 N east and a force of 11 N west act on a moving object. Model these forces by starting at X and drawing a red arrow three spaces to the right. Then, start at that point and draw a blue arrow 11 spaces to the left. The net force is modeled by how far this point is from X, 8 N west.

3. Are the forces you modeled balanced or unbalanced? Will the forces change the object's motion?

## Apply It

4. Suppose a force of 8 N east, a force of 12 N west, and a force of 4 N east act on a moving object. Use different colors of markers to model the forces on the object.

5. What is the net force on the object? Are the forces you modeled balanced or unbalanced? Will the forces change the object's motion?

6. **Model** other examples of balanced and unbalanced forces acting on an object. In each case, decide which forces will act on the object.

7. 🔑 **Key Concept** For each of the forces you modeled, determine the net force, and decide if the forces are balanced or unbalanced. Then, decide if the forces will change the object's motion.

Net Force 8 N West

# Newton's Second Law

### Reading Guide

**Key Concepts**
ESSENTIAL QUESTIONS

- What is Newton's second law of motion?
- How does centripetal force affect circular motion?

**Vocabulary**

**Newton's second law of motion** p. 65

**circular motion** p. 66

**centripetal force** p. 66

**g** Multilingual eGlossary

**Inquiry** **What makes it go?**

The archer pulls back the string and takes aim. When she releases the string, the arrow soars through the air. To reach the target, the arrow must quickly reach a high speed. How is it able to move so fast? The force from the string determines the arrow's speed.

## Launch Lab

### What forces affect motion along a curved path?

When traveling in a car or riding on a roller coaster, you can feel different forces acting on you as you move along a curved path. What are these forces? How do they affect your motion?

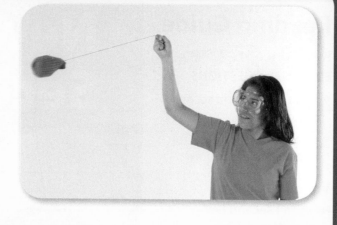

1 Read and complete a lab safety form.

2 Attach a piece of **string** about 1 m long to a rolled-up **sock**.

**WARNING:** Find a spot away from your classmates for the next steps.

3 While holding the end of the string, swing the sock around in a circle above your head. Notice the force tugging on the string.

4 Repeat step 3 with two socks rolled together. In your Science Journal, compare the force of swinging one sock to the force of swinging two socks.

#### Think About This

1. Describe the forces acting along the string while you were swinging it. Classify each force as balanced or unbalanced.

2. 🔑 **Key Concept** How does the force from the string seem to affect the sock's motion?

## How do forces change motion?

Think about different ways that forces can change an object's motion. For example, how do forces change the motion of someone riding a bicycle? The forces of the person's feet on the pedals cause the wheels of the bicycle to turn faster and the bicycle's speed to increase. The speed of a skater slowly sliding across ice gradually decreases because of friction between the skates and the ice. Suppose you are pushing a wheelbarrow across a yard. You can change its speed by pushing with more or less force. You can change its direction by pushing it in the direction you want to move. Forces change an object's motion by changing its speed, its direction, or both its speed and its direction.

## Unbalanced Forces and Velocity

Velocity is speed in a certain direction. Only unbalanced forces change an object's velocity. A bicycle's speed will not increase unless the forces of the person's feet on the pedals is greater than friction that slows the wheels. A skater's speed will not decrease if the skater pushes back against the ice with a force greater than the friction against the skates. If someone pushes the wheelbarrow with the same force but in the opposite direction that you are pushing, the wheelbarrow's direction will not change.

In the previous lesson, you read about Newton's first law of motion—balanced forces do not change an object's velocity. In this lesson you will read about how unbalanced forces affect the velocity of an object.

## Unbalanced Forces on an Object at Rest

An example of how unbalanced forces affect an object at rest is shown in **Figure 16.** At first the ball is not moving. The hand holds the ball up against the downward pull of gravity. Because the forces on the ball are balanced, the ball remains at rest. When the hand moves out of the way, the ball falls downward. You know that the forces on the ball are now unbalanced because the ball's motion changed. The ball moves in the direction of the net force. When unbalanced forces act on an object at rest, the object begins moving in the direction of the net force.

## Unbalanced Forces on an Object in Motion

Unbalanced forces change the velocity of a moving object. Recall that one way to change an object's velocity is to change its speed.

**Speeding Up** If the net force acting on a moving object is in the direction that the object is moving, the object will speed up. For example, a net force acts on the sled in **Figure 17.** Because the net force is in the direction of motion, the sled's speed increases.

**Slowing Down** Think about what happens if the direction of the net force on an object is opposite to the direction the object moves. The object slows down. When the boy sliding on the sled in **Figure 17** pushes his foot against the snow, friction acts in the direction opposite to his motion. Because the net force is in the direction opposite to the sled's motion, the sled's speed decreases.

**Reading Check** What happens to the speed of a wagon rolling to the right if a net force to the right acts on it?

**Balanced Forces**

Force exerted by hand

Force due to gravity

**Unbalanced Forces**

Force due to gravity

▲ **Figure 16** When unbalanced forces act on a ball at rest, it moves in the direction of the net force.

**Speeding up**

Velocity

Net force

**Slowing down**

Velocity

Net force

◀ **Figure 17** Unbalanced forces can cause an object to speed up or slow down.

**Visual Check** How would the net force and velocity arrows in the left photo change if the girl pushed harder?

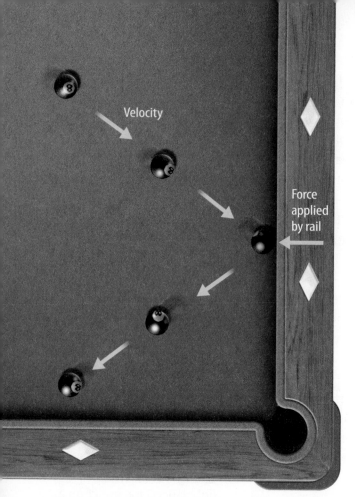

Velocity

Force applied by rail

**Figure 18** Unbalanced forces act on the billiard ball, causing its direction to change.

## Changes in Direction of Motion

Another way that unbalanced forces can change an object's velocity is to change its direction. The ball in **Figure 18** moved at a constant velocity until it hit the rail of the billiard table. The force applied by the rail changed the ball's direction. Likewise, unbalanced forces change the direction of Earth's crust. Recall that the crust is broken into moving pieces called plates. The direction of one plate changes when another plate pushes against it with an unbalanced force.

## Unbalanced Forces and Acceleration

You have read how unbalanced forces can change an object's velocity by changing its speed, its direction, or both its speed and its direction. Another name for a change in velocity over time is acceleration. When the girl in **Figure 17** pushed the sled, the sled accelerated because its speed changed. When the billiard ball in **Figure 18** hit the side of the table, the ball accelerated because its direction changed. Unbalanced forces can make an object accelerate by changing its speed, its direction, or both.

 **Reading Check** How do unbalanced forces affect an object at rest or in motion?

---

**Inquiry MiniLab**

**10 minutes**

### How are force and mass related? 👓 ✋

Unbalanced forces cause an object to accelerate. If the mass of the object increases, how does the force required to accelerate the object change?

1. Read and complete a lab safety form.
2. Tie a **string** to a **small box.** Pull the box about 2 m across the floor. Notice the force required to cause the box to accelerate.
3. Put **clay** in the box to increase its mass. Pull the box so that its acceleration is about the same as before. Notice the force required.

**Analyze and Conclude**

1. **Compare** the strength of the force needed to accelerate the box each time.

2. 🔑 **Key Concept** How did the mass affect the force needed to accelerate the box?

# Newton's Second Law of Motion

Isaac Newton also described the relationship between an object's acceleration (change in velocity over time) and the net force that acts on an object. *According to* **Newton's second law of motion,** *the acceleration of an object is equal to the net force acting on the object divided by the object's mass.* The direction of acceleration is the same as the direction of the net force.

 **Key Concept Check** What is Newton's second law of motion?

### Newton's Second Law Equation

$$\text{acceleration (in m/s}^2) = \frac{\text{net force (in N)}}{\text{mass (in kg)}}$$

$$a = \frac{F}{m}$$

Notice that the equation for Newton's second law has SI units. Acceleration is expressed in meters per second squared (m/s$^2$), mass in kilograms (kg), and force in newtons (N). From this equation, it follows that a newton is the same as kg·m/s$^2$.

**FOLDABLES**

Make a half-book from a sheet of notebook paper. Use it to organize your notes on Newton's second law.

Newton's Second Law

---

## Math Skills — Newton's Second Law Equation

**Solve for Acceleration** You throw a 0.5-kg basketball with a force of 10 N. What is the acceleration of the ball?

**1** **This is what you know:**

mass:      $m = 0.5$ kg

force:      $F = 10$ N or 10 kg·m/s$^2$

**2** **This is what you need to find:**

acceleration:      $a$

**3** **Use this formula:**

$$a = \frac{F}{m}$$

**4** **Substitute:**
the values for **F** and **m** into the formula and divide.

$$a = \frac{10\text{ N}}{0.5\text{ kg}} = 20\ \frac{\text{kg·m/s}^2}{\text{kg}} = 20\text{ m/s}^2$$

**Answer:** The acceleration of the ball is 20 m/s$^2$.

### Practice

1. A 24-N net force acts on an 8-kg rock. What is the acceleration of the rock?

2. A 30-N net force on a skater produces an acceleration of 0.6 m/s$^2$. What is the mass of the skater?

3. What net force acting on a 14-kg wagon produces an acceleration of 1.5 m/s$^2$?

**Review**
- Math Practice
- Personal Tutor

# Circular Motion

Newton's second law of motion describes the relationship between an object's change in velocity over time, or acceleration, and unbalanced forces acting on the object. You already read how this relationship applies to motion along a line. It also applies to circular motion. **Circular motion** *is any motion in which an object is moving along a curved path.*

## Centripetal Force

The ball in **Figure 19** is in circular motion. The velocity arrows show that the ball has a tendency to move along a straight path. Inertia—not a force—causes this motion. The ball's path is curved because the string pulls the ball inward. *In circular motion, a force that acts perpendicular to the direction of motion, toward the center of the curve, is* **centripetal** (sen TRIH puh tuhl) **force.** The figure also shows that the ball accelerates in the direction of the centripetal force.

 **Key Concept Check** How does centripetal force affect circular motion?

## The Motion of Satellites and Planets

Another object that experiences centripetal force is a satellite. A satellite is any object in space that orbits a larger object. Like the ball in **Figure 19,** a satellite tends to move along a straight path because of inertia. But just as the string pulls the ball inward, gravity pulls a satellite inward. Gravity is the centripetal force that keeps a satellite in orbit by changing its direction. The Moon is a satellite of Earth. As shown in **Figure 19,** Earth's gravity changes the Moon's direction. Similarly, the Sun's gravity changes the direction of its satellites, including Earth.

**WORD ORIGIN** ············

centripetal
from Latin *centripetus*, means "toward the center"

**Figure 19** Inertia of the moving object and the centripetal force acting on the object produce the circular motion of the ball and the Moon.

**Visual Check** How does the direction of the velocity of a satellite differ from the direction of its acceleration?

**Circular Motion**

**Concepts in Motion** Animation

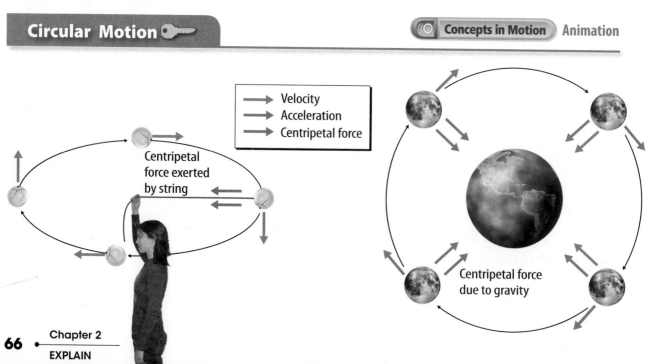

Velocity
Acceleration
Centripetal force

Centripetal force exerted by string

Centripetal force due to gravity

# Lesson 3 Review

## Visual Summary

Unbalanced forces cause an object to speed up, slow down, or change direction.

Newton's second law of motion relates an object's acceleration to its mass and the net force on the object.

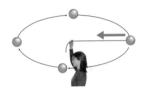

Any motion in which an object is moving along a curved path is circular motion.

**FOLDABLES**

Use your lesson Foldable to review the lesson. Save your Foldable for the project at the end of the chapter.

## What do you think NOW?

You first read the statements below at the beginning of the chapter.

**5.** When an object's speed increases, the object accelerates.

**6.** If an object's mass increases, its acceleration also increases if the net force acting on the object stays the same.

Did you change your mind about whether you agree or disagree with the statements? Rewrite any false statements to make them true.

## Use Vocabulary

**1** **Explain** Newton's second law of motion in your own words.

**2** **Use the term** *circular motion* in a sentence.

## Understand Key Concepts

**3** A cat pushes a 0.25-kg toy with a net force of 8 N. According to Newton's second law what is the acceleration of the ball?
  **A.** 2 m/s$^2$      **C.** 16 m/s$^2$
  **B.** 4 m/s$^2$      **D.** 32 m/s$^2$

**4** **Describe** how centripetal force affects circular motion.

## Interpret Graphics

**5** **Apply** Copy and fill in the graphic organizer below. Give examples of unbalanced forces on an object that could cause an object to accelerate.

**6** **Complete** Copy the graphic organizer below and complete each equation according to Newton's second law.

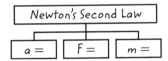

## Critical Thinking

**7** **Design** You need to lift up a 45-N object. Draw an illustration that explains the strength and direction of the force used to lift the object.

## Math Skills ✕÷      **Review**
─── Math Practice ───

**8** The force of Earth's gravity is about 10 N downward. What is the acceleration of a 15-kg backpack if lifted with a 15-N force?

# How does a change in mass or force affect acceleration?

## Materials

baseball

foam ball

meterstick

## Safety

Force, mass, and acceleration are all related variables. In this activity, you will use these variables to study Newton's second law of motion.

### Learn It

*Vary* means "to change." A **variable** is a quantity that can be changed. For example, the variables related to Newton's second law of motion are force, mass, and acceleration. You can find the relationship between any two of these variables by changing one of them and keeping the third variable the same.

### Try It

1. Read and complete a lab safety form.

2. Hold a baseball in one hand and a foam ball in your other hand. Compare the masses of the two balls.

3. Lay both balls on a flat surface. Push a meterstick against the balls at the same time with the same force. Compare the accelerations of the ball.

4. Using only the baseball and the meterstick, lightly push the ball and observe its acceleration. Again observe the acceleration as you push the baseball with a stronger push. Compare the accelerations of the ball when you used a weak force and when you used a strong force.

### Apply It

5. Answer the following questions for both step 3 and step 4. What variable did you change? What variable changed as a result? What variable did you keep the same?

6. Using your results, state the relationship between acceleration and mass if the net force on an object does not change. Then, state the relationship between acceleration and force if mass does not change.

7. 🔑 **Key Concept** How do your results support Newton's second law of motion?

# Newton's Third Law

## Reading Guide

### Key Concepts
**ESSENTIAL QUESTIONS**

- What is Newton's third law of motion?
- Why don't the forces in a force pair cancel each other?
- What is the law of conservation of momentum?

### Vocabulary

**Newton's third law of motion** p. 71

**force pair** p. 71

**momentum** p. 73

**law of conservation of momentum** p. 74

**g** **Multilingual eGlossary**

---

**inquiry** **Why move up?**

To reach the height she needs for her dive, this diver must move up into the air. Does she just jump up? No, she doesn't. She pushes down on the diving board and the diving board propels her into the air. How does pushing down cause the diver to move up?

## How do opposite forces compare?

If you think about forces you encounter every day, you might notice forces that occur in pairs. For example, if you drop a rubber ball, the falling ball pushes against the floor. The ball bounces because the floor pushes with an opposite force against the ball. How do these opposite forces compare?

1. Read and complete a lab safety form.
2. Stand so that you face your lab partner, about half a meter away. Each of you should hold a **spring scale.**
3. Hook the two scales together, and gently pull them away from each other. Notice the force reading on each scale.
4. Pull harder on the scales, and again notice the force readings on the scales.
5. Continue to pull on both scales, but let the scales slowly move toward your lab partner and then toward you at a constant speed.

**Think About This**

1. Identify the directions of the forces on each scale. Record this information in your Science Journal.

2. 🔑 **Key Concept** Describe the relationship you noticed between the force readings on the two scales.

**Figure 20** When the skater pushes against the wall, the wall applies a force to the skater that pushes him away from the wall.

## Opposite Forces

Have you ever been on in-line skates and pushed against a wall? When you pushed against the wall, like the boy is doing in **Figure 20,** you started moving away from it. What force caused you to move?

You might think the force of the muscles in your hands moved you away from the wall. But think about the direction of your push. You pushed against the wall in the opposite direction from your movement. It might be hard to imagine, but when you pushed against the wall, the wall pushed back in the opposite direction. The push of the wall caused you to accelerate away from the wall. When an object applies a force on another object, the second object applies a force of the same strength on the first object, but the force is in the opposite direction.

✓ **Reading Check** When you are standing, you push on the floor, and the floor pushes on you. How do the directions and strengths of these forces compare?

# Newton's Third Law of Motion

Newton's first two laws of motion describe the effects of balanced and unbalanced forces on one object. Newton's third law relates forces between two objects. *According to* **Newton's third law of motion,** *when one object exerts a force on a second object, the second object exerts an equal force in the opposite direction on the first object.* An example of forces described by Newton's third law is shown in **Figure 21.** When the gymnast pushes against the vault, the vault pushes back against the gymnast. Notice that the lengths of the force arrows are the same, but the directions are opposite.

 **Key Concept Check** What is Newton's third law of motion?

## Force Pairs

The forces described by Newton's third law depend on each other. *A* **force pair** *is the forces two objects apply to each other.* Recall that you can add forces to calculate the net force. If the forces of a force pair always act in opposite directions and are always the same strength, why don't they cancel each other? The answer is that each force acts on a different object. In **Figure 22,** the girl's feet act on the boat. The force of the boat acts on the girl's feet. The forces do not result in a net force of zero because they act on different objects. Adding forces can only result in a net force of zero if the forces act on the same object.

 **Key Concept Check** Why don't the forces in a force pair cancel each other?

## Action and Reaction

In a force pair, one force is called the action force. The other force is called the reaction force. The girl in **Figure 22** applies an action force against the boat. The reaction force is the force that the boat applies to the girl. For every action force, there is a reaction force that is equal in strength but opposite in direction.

Newton's Third Law

Force exerted on the vault

Force exerted on the gymnast

▲ **Figure 21** The force of the vault propels the gymnast upward.

Force Pairs

Action force exerted by the girl on the boat

Reaction force exerted by the boat on the girl

▲ **Figure 22** The force pair is the force the girl applies to the boat and the force that the boat applies to the girl.

**Visual Check** How can you tell that the forces don't cancel each other?

**FOLDABLES®**

Make a half-book from a sheet of notebook paper. Use it to summarize how Newton's third law explains the motion of a variety of common activities.

*Using Newton's Third Law*

# Using Newton's Third Law of Motion

When you push against an object, the force you apply is called the action force. The object then pushes back against you. The force applied by the object is called the reaction force. According to Newton's second law of motion, when the reaction force results in an unbalanced force, there is a net force, and the object accelerates. As shown in **Figure 23**, Newton's third law explains how you can swim and jump. It also explains how rockets can be launched into space.

✓ **Reading Check** How does Newton's third law apply to the motion of a bouncing ball?

## Action and Reaction Forces 🔑

**Figure 23** Every action force has a reaction force in the opposite direction.

◀ **Swimming** When you swim, you push your arms against the water in the pool. The water in the pool pushes back on you in the opposite (forward) direction. If your arms push the water back with enough force, the reaction force of the water on your body is greater than the force of fluid friction. The net force is forward. You accelerate in the direction of the net force and swim forward through the water.

▶ **Jumping** When you jump, you push down on the ground, and the ground pushes up on you. The upward force of the ground combines with the downward force of gravity to form the net force acting on you. If you push down hard enough, the upward reaction force is greater than the downward force of gravity. The net force is upward. According to Newton's second law, your acceleration is in the same direction as the net force, so you accelerate upward.

◀ **Rocket Motion** The burning fuel in a rocket engine produces a hot gas. The engine pushes the hot gas out in a downward direction. The gas pushes upward on the engine. When the upward force of the gas pushing on the engine becomes greater than the downward force of gravity on the rocket, the net force is in the upward direction. The rocket then accelerates upward.

✓ **Visual Check** On what part of the swimmer's body does the water's reaction force push?

# Momentum

Because action and reaction forces do not cancel each other, they can change the motion of objects. A useful way to describe changes in velocity is by describing momentum. **Momentum** *is a measure of how hard it is to stop a moving object.* It is the product of an object's mass and velocity. An object's momentum is in the same direction as its velocity.

**WORD ORIGIN** ············

**momentum**
from Latin *momentum*, means "movement, impulse"

## Momentum Equation

momentum (in kg·m/s) = mass (in kg) × velocity (in m/s)

$$p = m \times v$$

If a large truck and a car move at the same speed, the truck is harder to stop. Because it has more mass, it has more momentum. If cars of equal mass move at different speeds, the faster car has more momentum and is harder to stop.

Newton's first two laws relate to momentum. According to Newton's first law, if the net force on an object is zero, its velocity does not change. This means its momentum does not change. Newton's second law states that the net force on an object is the product of its mass and its change in velocity. Because momentum is the product of mass and velocity, the force on an object equals its change in momentum.

## Math Skills    Finding Momentum

**Solve for Momentum** What is the momentum of a 12-kg bicycle moving at 5.5 m/s?

**1 This is what you know:**     mass:      $m = 12$ kg
                                 velocity:  $v = 5.5$ m/s

**2 This is what you need to find:**     momentum:   $p$

**3 Use this formula:**     $p = m \times v$

**4 Substitute:**     $p = 12$ kg $\times$ 5.5 m/s = 66 kg·m/s
the values for *m* and *v*
into the formula and multiply.

**Answer:** The momentum of the bicycle is 66 kg·m/s in the direction of the velocity.

**Review**
• Math Practice
• Personal Tutor

## Practice

1. What is the momentum of a 1.5-kg ball rolling at 3.0 m/s?

2. A 55-kg woman has a momentum of 220 kg·m/s. What is her velocity?

## Conservation of Momentum

You might have noticed that if a moving ball hits another ball that is not moving, the motion of both balls changes. The cue ball in **Figure 24** has momentum because it has mass and is moving. When it hits the other balls, the cue ball's velocity and momentum decrease. Now the other balls start moving. Because these balls have mass and velocity, they also have momentum.

### The Law of Conservation of Momentum

In any collision, one object transfers momentum to another object. The billiard balls in **Figure 24** gain the momentum lost by the cue ball. The total momentum, however, does not change. *According to the* **law of conservation of momentum,** *the total momentum of a group of objects stays the same unless outside forces act on the objects.* Outside forces include friction. Friction between the balls and the billiard table decreases their velocities, and they lose momentum.

 **Key Concept Check** What is the law of conservation of momentum?

### Types of Collisions

Objects collide with each other in different ways. When colliding objects bounce off each other, it is an elastic collision. If objects collide and stick together, such as when one football player tackles another, the collision is inelastic. No matter the type of collision, the total momentum will be the same before and after the collision.

**Figure 24** The total momentum of all the balls is the same before and after the collision.

---

(Inquiry) **MiniLab**                                    15 minutes

### Is momentum conserved during a collision?

1 Read and complete a lab safety form.

2 Make a track by using **masking tape** to secure two **metersticks** side by side on a table, about 4 cm apart.

3 Place two **tennis balls** on the track. Roll one ball against the other. Then, roll the balls at about the same speed toward each other.

4 Place the balls so that they touch. Observe the collision as you gently roll **another ball** against them.

**Analyze and Conclude**

1. **Explain** how you know that momentum was transferred from one ball to another.

2.  **Key Concept** What could you measure to show that momentum is conserved?

# Lesson 4 Review

## Visual Summary

Newton's third law of motion describes the force pair between two objects.

For every action force, there is a reaction force that is equal in strength but opposite in direction.

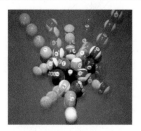

In any collision, momentum is transferred from one object to another.

**FOLDABLES**

Use your lesson Foldable to review the lesson. Save your Foldable for the project at the end of the chapter.

## What do you think NOW?

You first read the statements below at the beginning of the chapter.

**7.** If objects collide, the object with more mass applies more force.

**8.** Momentum is a measure of how hard it is to stop a moving object.

Did you change your mind about whether you agree or disagree with the statements? Rewrite any false statements to make them true.

## Use Vocabulary

1 **Define** *momentum* in your own words.

2 The force of a bat on a ball and the force of a ball on a bat are a(n) _____.

## Understand Key Concepts 🔑

3 **State** Newton's third law of motion.

4 A ball with momentum 16 kg·m/s strikes a ball at rest. What is the total momentum of both balls after the collision?
   A. −16 kg·m/s          C. 8 kg·m/s
   B. −8 kg·m/s           D. 16 kg·m/s

5 **Identify** A child jumps on a trampoline. The trampoline bounces her up. Why don't the forces cancel each other?

## Interpret Graphics

6 **Predict** what will happen to the velocity and momentum of each ball when the small ball hits the heavier large ball?

7 **Organize** Copy and fill in the table.

| Event | Action Force | Reaction Force |
|---|---|---|
| A girl kicks a soccer ball. |  |  |
| A book sits on a table. |  |  |

## Critical Thinking

8 **Decide** How is it possible for a bicycle to have more momentum than a truck?

## Math Skills ×÷+−

 **Review**
— Math Practice —

9 A 2.0-kg ball rolls to the right at 3.0 m/s. A 4.0-kg ball rolls to the left at 2.0 m/s. What is the momentum of the system after a head-on collision of the two balls?

## Materials

plastic lid

golf ball

modeling clay

2.5-N spring scales (2)

### Safety

# Modeling Newton's Laws of Motion

Newton's first and second laws of motion describe the relationship between unbalanced forces and motion. These laws relate to forces acting on one object. Newton's third law describes the strength and direction of force pairs. This law relates to forces on two different objects. You can learn about all three of Newton's laws of motion by modeling them.

## Question

How can you model Newton's laws of motion?

## Procedure

1. Read and complete a lab safety form.

2. Attach a spring scale to a plastic lid. Add mass to the lid by placing a ball of modeling clay on it.

3. Slowly pull the lid along a table with the spring scale. Record the force reading on the scale in your Science Journal.

4. Try to use the spring scale to pull the lid with constant force and constant speed.

5. Try pulling the lid with increasing force and constant speed.

6. Pull the lid so that it accelerates quickly.

7. Increase the mass of the lid by adding more modeling clay. Repeat steps 3–6.

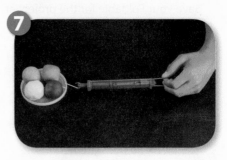

8. Replace the modeling clay with a golf ball. Try pulling the lid slowly. Then, try pulling it from a standstill quickly. What happens to the ball in each case? Record your results.

**9** To model Newton's first law of motion, design an activity using the lid and the spring scales that shows that a net force of zero does not change the motion of an object.

**10** To model Newton's second law of motion, design an activity that shows that if the net force acting on an object is not zero, the object accelerates.

**11** To model Newton's third law of motion, plan an activity that shows action and reaction forces on an object.

**12** After your teacher approves your plan, perform the activities.

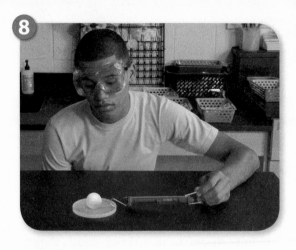

## Analyze and Conclude

**13** **Identify** the variables in each of your models. Which variables changed and which remained constant?

**14** 🔵 **The Big Idea** For each law of motion that you modeled, how did the force applied to the lid relate to the motion of the lid?

## Communicate Your Results

Choose one of the laws of motion, and model it for the class. Compare your model with the method of modeling used by other lab groups.

 **Extension**

Describe another way you could model Newton's three laws of motion using materials other than those used in this lab. For example, for Newton's first law of motion, you could pedal a bicycle at a constant speed.

**Lab Tips**

☑ Use a smooth surface so that the lid moves easily.

☑ You might want to make a data table in which you can record your observations and the force readings.

**Remember** to use scientific methods.

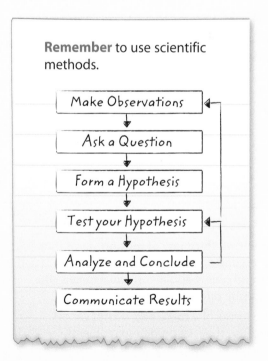

Make Observations

Ask a Question

Form a Hypothesis

Test your Hypothesis

Analyze and Conclude

Communicate Results

# Chapter 2 Study Guide

**THE BIG IDEA** An object's motion changes if a net force acts on the object.

| Key Concepts Summary 🔑 | Vocabulary |
|---|---|
| **Lesson 1: Gravity and Friction**<br>• Friction is a **contact force.** Magnetism is a **noncontact force.**<br>• The law of universal gravitation states that all objects are attracted to each other by **gravity.**<br>• **Friction** can stop or slow down objects sliding past each other.<br> | **force** p. 45<br>**contact force** p. 45<br>**noncontact force** p. 46<br>**gravity** p. 47<br>**mass** p. 47<br>**weight** p. 48<br>**friction** p. 49 |
| **Lesson 2: Newton's First Law**<br>• An object's motion can only be changed by **unbalanced forces.**<br>• According to **Newton's first law of motion,** the motion of an object is not changed by **balanced forces** acting on it.<br>• **Inertia** is the tendency of an object to resist a change in its motion.<br> | **net force** p. 55<br>**balanced forces** p. 56<br>**unbalanced forces** p. 56<br>**Newton's first law of motion** p. 57<br>**inertia** p. 58 |
| **Lesson 3: Newton's Second Law**<br>• According to **Newton's second law of motion,** an object's acceleration is the net force on the object divided by its mass.<br>• In **circular motion,** a **centripetal force** pulls an object toward the center of the curve.<br> | **Newton's second law of motion** p. 65<br>**circular motion** p. 66<br>**centripetal force** p. 66 |
| **Lesson 4: Newton's Third Law**<br>• **Newton's third law of motion** states that when one object applies a force on another, the second object applies an equal force in the opposite direction on the first object.<br>• The forces of a **force pair** do not cancel because they act on different objects.<br>• According to the **law of conservation of momentum,** momentum is conserved during a collision unless an outside force acts on the colliding objects.<br> | **Newton's third law of motion** p. 71<br>**force pair** p. 71<br>**momentum** p. 73<br>**law of conservation of momentum** p. 74 |

## FOLDABLES® Chapter Project

Ass emble your lesson Foldables as shown to make a Chapter Project. Use the project to review what you have learned in this chapter.

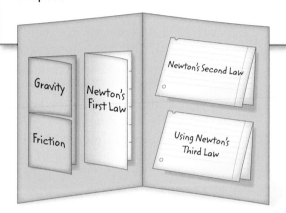

## Use Vocabulary

1 Kilograms is the SI unit for measuring _____.

2 The force of gravity on an object is its _____.

3 The sum of all the forces on an object is the _____.

4 An object that has _____ acting on it acts as if there were no forces acting on it at all.

5 A car races around a circular track. Friction on the tires is the _____ that acts toward the center of the circle and keeps the car on the circular path.

6 A heavy train requires nearly a mile to come to a complete stop because it has a lot of _____.

## Link Vocabulary and Key Concepts

((◎ Concepts in Motion)) Interactive Concept Map

*Copy this concept map, and then use vocabulary terms from the previous page to complete the concept map.*

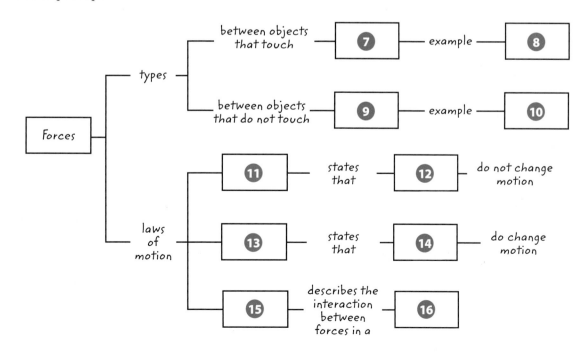

# Chapter 2 Review

## Understand Key Concepts

**1** The arrows in the figure represent the gravitational force between marbles that have equal mass.

How should the force arrows look if a marble that has greater mass replaces one of these marbles?

A. Both arrows should be drawn longer.
B. Both arrows should stay the same length.
C. The arrow from the marble with less mass should be longer than the other arrow.
D. The arrow from the marble with less mass should be shorter than the other arrow.

**2** A person pushes a box across a flat surface with a force less than the weight of the box. Which force is weakest?

A. the force of gravity on the box
B. the force of the table on the box
C. the applied force against the box
D. the sliding friction against the box

**3** A train moves at a constant speed on a straight track. Which statement is true?

A. No horizontal forces act on the train as it moves.
B. The train moves only because of its inertia.
C. The forces of the train's engine balances friction.
D. An unbalanced force keeps the train moving.

**4** The Moon orbits Earth in a nearly circular orbit. What is the centripetal force?

A. the pull of the Moon on Earth
B. the outward force on the Moon
C. the Moon's inertia as it orbits Earth
D. Earth's gravitational pull on the Moon

**5** A 30-kg television sits on a table. The acceleration due to gravity is 10 m/s$^2$. What force does the table exert on the television?

A. 0.3 N
B. 3 N
C. 300 N
D. 600 N

**6** Which does NOT describe a force pair?

A. When you push on a bike's brakes, the friction between the tires and the road increases.
B. When a diver jumps off a diving board, the board pushes the diver up.
C. When an ice skater pushes off a wall, the wall pushes the skater away from the wall.
D. When a boy pulls a toy wagon, the wagon pulls back on the boy.

**7** A box on a table has these forces acting on it.

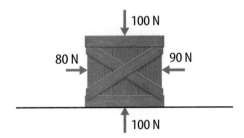

What is the static friction between the box and the table?

A. 0 N
B. 10 N
C. greater than 10 N
D. between 0 and 10 N

**8** A 4-kg goose swims with a velocity of 1 m/s. What is its momentum?

A. 4 N
B. 4 kg·m/s$^2$
C. 4 kg·m/s
D. 4 m/s$^2$

## Critical Thinking

**9** **Predict** If an astronaut moved away from Earth in the direction of the Moon, how would the gravitational force between Earth and the astronaut change? How would the gravitational force between the Moon and the astronaut change?

**10** **Analyze** A box is on a table. Two people push on the box from opposite sides. Which of the labeled forces make up a force pair? Explain your answer.

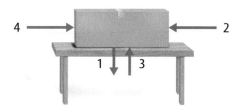

**11** **Conclude** A refrigerator has a maximum static friction force of 250 N. Sam can push the refrigerator with a force of 130 N. Amir and André can each push with a force of 65 N. How could they all move the refrigerator? Will the refrigerator move with constant velocity? Why or why not?

**12** **Give an example** of unbalanced forces acting on an object.

**13** **Infer** Two skaters stand on ice. One weighs 250 N, and the other weighs 500 N. They push against each other and move in opposite directions. Which one will travel farther before stopping? Explain your answer.

*Writing in Science*

**14** Imagine that you are an auto designer. Your job is to design brakes for different automobiles. Write a four-sentence plan that explains what you need to consider about momentum when designing brakes for a heavy truck, a light truck, a small car, and a van.

## REVIEW THE BIG IDEA

**15** Explain how balanced and unbalanced forces affect objects that are not moving and those that are moving.

**16** The photo below shows people on a carnival swing ride. How do forces change the motion of the riders?

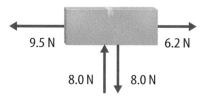

## Math Skills

### Solve One-Step Equations

**17** A net force of 17 N is applied to an object, giving it an acceleration of 2.5 m/s². What is the mass of the object?

**18** A tennis ball's mass is about 0.60 kg. Its velocity is 2.5 m/s. What is the momentum of the ball?

**19** A box with a mass of 0.82 kg has these forces acting on it.

9.5 N          6.2 N

8.0 N     8.0 N

What is the strength and direction of the acceleration of the box?

# Standardized Test Practice

*Record your answers on the answer sheet provided by your teacher or on a sheet of paper.*

## Multiple Choice

**1** A baseball has an approximate mass of 0.15 kg. If a bat strikes the baseball with a force of 6 N, what is the acceleration of the ball?

**A** 4 m/s²

**B** 6 m/s²

**C** 40 m/s²

**D** 60 m/s²

*Use the diagram below to answer question 2.*

**2** The person in the diagram above is unable to move the crate from its position. Which is the opposing force?

**A** gravity

**B** normal force

**C** sliding friction

**D** static friction

**3** The mass of a person on Earth is 72 kg. What is the mass of the same person on the Moon where gravity is one-sixth that of Earth?

**A** 12 kg

**B** 60 kg

**C** 72 kg

**D** 432 kg

**4** A swimmer pushing off from the wall of a pool exerts a force of 1 N on the wall. What is the reaction force of the wall on the swimmer?

**A** 0 N

**B** 1 N

**C** 2 N

**D** 10 N

*Use the diagram below to answer questions 5 and 6.*

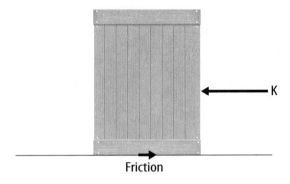

**5** Which term applies to the forces in the diagram above?

**A** negative

**B** positive

**C** reference

**D** unbalanced

**6** In the diagram above, what happens when force *K* is applied to the crate at rest?

**A** The crate remains at rest.

**B** The crate moves back and forth.

**C** The crate moves to the left.

**D** The crate moves to the right.

**7** What is another term for change in velocity?

**A** acceleration

**B** inertia

**C** centripetal force

**D** maximum speed

*Use the diagram below to answer question 8.*

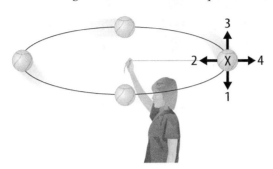

**8** The person in the diagram is spinning a ball on a string. When the ball is in position *X*, what is the direction of the centripetal force?

**A** 1

**B** 2

**C** 3

**D** 4

**9** Which is ALWAYS a contact force?

**A** electric

**B** friction

**C** gravity

**D** magnetic

**10** When two billiard balls collide, which is ALWAYS conserved?

**A** acceleration

**B** direction

**C** force

**D** momentum

## Constructed Response

*Use the table below to answer question 11.*

| Newton's Laws of Motion | Explanation |
|---|---|
| First | |
| Second | |
| Third | |

**11** Explain each of Newton's laws of motion. What is one practical application of each law?

*Use the diagram below to answer questions 12 and 13.*

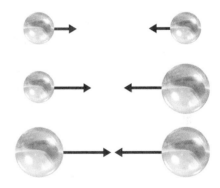

**12** The arrows in the diagram above represent forces. What scientific law does the diagram illustrate? What does the law state?

**13** Using the diagram, explain how marble mass affects gravitational attraction.

| NEED EXTRA HELP? | | | | | | | | | | | | | |
|---|---|---|---|---|---|---|---|---|---|---|---|---|---|
| If You Missed Question... | 1 | 2 | 3 | 4 | 5 | 6 | 7 | 8 | 9 | 10 | 11 | 12 | 13 |
| Go to Lesson... | 3 | 1 | 1 | 4 | 2 | 2 | 3 | 3 | 1 | 4 | 2–4 | 1 | 1 |

# Work and Simple Machines

**THE BIG IDEA**

### How do machines make doing work easier?

## Inquiry  Hard Work or Not?

Digging this hole with a hand shovel would be hard work. Using an earthmover makes the task easier to do. You probably think writing an essay for English class is hard work, but is it?

- What do you think work is?

- How do you think work and energy are related?

- How do you think machines make doing work easier?

# Get Ready to Read

## What do you think?

Before you read, decide if you agree or disagree with each of these statements. As you read this chapter, see if you change your mind about any of the statements.

**1** You do work when you push a book across a table.

**2** Doing work faster requires more power.

**3** Machines always decrease the force needed to do a job.

**4** A well-oiled, low-friction machine can be 100 percent efficient.

**5** A doorknob is a simple machine.

**6** A loading ramp makes it easier to lift a load.

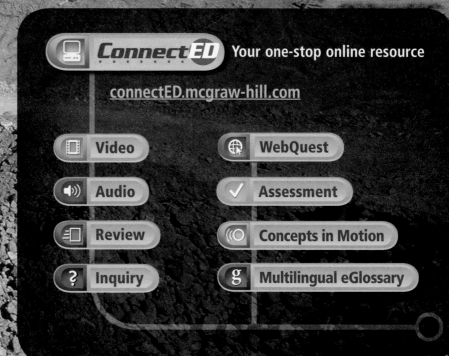

**ConnectED** Your one-stop online resource

connectED.mcgraw-hill.com

- Video
- Audio
- Review
- Inquiry
- WebQuest
- Assessment
- Concepts in Motion
- Multilingual eGlossary

# Work and Power

## Reading Guide

### Key Concepts 🔑
**ESSENTIAL QUESTIONS**

- What must happen for work to be done?
- How does doing work on an object change its energy?
- How are work and power related?

### Vocabulary
**work** p. 87

**power** p. 91

 **Multilingual eGlossary**

**Inquiry** **Enough Power?**

A powerful tugboat can pull this large cargo ship. Could a much smaller boat also get this work done? As you will soon read, work and power are closely related.

### How do you know when work is done?

Many things that seem like hard work to you are not work at all in the scientific sense. This activity explores the scientific meaning of work.

1. Read and complete a lab safety form.

2. Tie a **string** around a **book.** Center the string on the book, and place the book on a desk.

3. Attach a **spring scale** to the string. Have a partner read the scale as you pull on the scale as shown in the photo. Pull very gently at first and then slowly increase how hard you pull.

4. Record in your Science Journal the scale reading when the book begins to move.

**Think About This**

1. What amount of force was needed to make the book move? How do you know?

2. 🔑 **Key Concept** How do you think force and motion might be related to doing work? Explain your reasoning.

## What is work?

Students might say studying and doing homework is hard **work.** To a scientist, these things take no work at all. Why? Because the scientific meaning of work includes forces and motion, not concentration. In science, **work** is *the transfer of energy that occurs when a force is applied over a distance.*

Imagine pushing your bicycle into a bike rack. Your push (a force) makes your bicycle (an object) move. Therefore, work is done. You do no work if you push on the bicycle and it does not move. A force that does not make the object move does no work. The weight lifter shown in **Figure 1** does work in one example but not in the other.

**WORD ORIGIN** · · · · · · · · · · ·

**work**
from Old English *weorc,* means "activity"

**Figure 1** The weight lifter does work only when he moves the weights.

The weight lifter does work on the weights when he exerts a force that makes the weights move.

Although the weight lifter exerts force, no work is done on the weights because the weights are not moving.

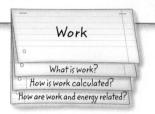

# Calculating Work

The amount of work done is easy to calculate. You must know two things to calculate work—force and distance, as shown in the following equation.

> ## Work Equation
> **Work** (in joules) = **force** (in newtons) × **distance** (in meters)
> $$W = Fd$$

The force must be in newtons (N) and distance must be in meters (m). When you multiply force and distance together, the result has units of newton-meter (N·m). The newton-meter is also known as the joule (J). Like other types of energy, work is measured in joules. The joule is the SI unit of work and energy.

The distance in the work equation is the distance the object moves while the force is acting on it. Suppose you push on a book over a distance of 0.25 m and then the book slides 3.0 m. Which distance do you use? You calculate the work done using 0.25 m because the force was applied along that distance.

 **Key Concept Check** How is work done?

---

**Math Skills** ✕ ➗ ➕ **Work Equation**

**Solve for Work** A student pushes a desk 2.0 m across the floor using a constant force of 25.0 N. How much work does the student do on the desk?

**1** **This is what you know:**
force: $F = 25.0$ N
distance: $d = 2.0$ m

**2** **This is what you need to find:** work: $W$

**3** **Use this formula:** $W = Fd$

**4** **Substitute:** values for $F$ and $d$ into the formula and multiply
$W = (25.0$ N$) \times (2.0$ m$) = 50.0$ N·m

**5** **Convert units:** (N) × (m) = N·m = J

**Answer:** The amount of work done is 50.0 J.

> **Review**
> • **Math Practice**
> • **Personal Tutor**

## Practice

A child pushes a toy truck 2.5 m across a floor with a constant force of 22 N. How much work does the child do on the toy truck?

Figure 2 A force that acts in the direction of the motion does work.

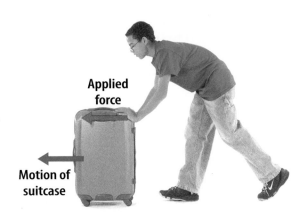

✓ **Visual Check** How are the applied forces different in each example?

## Factors That Affect Work

The work done on an object depends on the direction of the force applied and the direction of motion, as shown in **Figure 2**. Sometimes the force and the motion are in the same direction, such as when you push a suitcase along the floor. To calculate the work in this case, simply multiply the force and the distance.

**Force at an Angle** Now imagine pulling a wheeled suitcase, as shown in **Figure 2**. Note that the suitcase moves along the floor, but the force acts at an angle to the direction of the suitcase's motion. In other words, the force and the motion are not in the same direction. How do you calculate work?

Only the part of the force that acts in the direction of motion does work. Notice in **Figure 2** that the force has a horizontal part and a vertical part. Only the horizontal part of the force moves the suitcase across the floor. Therefore, only the horizontal part of the force is used in the work equation. The vertical part of the force does no work on the suitcase.

✓ **Reading Check** What part of a force does work when the direction of the force and the direction of motion are not the same?

---

Inquiry **MiniLab** 20 minutes

### What affects work? 🔧 ✋

Explore how different surfaces affect the work needed to pull an object across it.

1 Read and complete a lab safety form.

2 Tie a **string** around the center of a **book.** Attach a **spring scale** to the string.

3 Place the book on a desk next to a **meterstick.** Have a partner read the scale as you pull horizontally on it. Determine the least amount of force needed to slide the book 0.2 m at a constant speed. Record your data in your Science Journal in a table like the one below.

4 Repeat step 3 using the floor and other surfaces as instructed by your teacher.

| Sliding Book Data | | |
|---|---|---|
| Surface | Force (N) | Work (J) |
| Desk | | |
| Floor | | |

### Analyze and Conclude

1. **Determine** the work done in each trial.

2. **Compare** the surface requiring the least force to the one requiring the most force.

3. 🔑 **Key Concept** What caused the change in work done for each surface?

Upward force

Weight

▲ **Figure 3** The amount of force needed to lift the backpack is equal to the weight of the backpack.

**Lifting Objects** Lifting your backpack requires you to do work. The backpack has weight because of the downward force of gravity acting on it. To lift your backpack, you must pull upward with a force equal to or greater than the backpack's weight, as shown in **Figure 3.** The work done to lift any object is equal to the weight of the object multiplied by the distance it is lifted.

✓ **Reading Check** How do you calculate the work done when lifting an object?

## Work and Energy

Doing work on an object transfers energy to the object. This is important to scientists because it helps them predict how an object will act when forces are applied to it. Recall that moving objects have kinetic energy. The boy in **Figure 4** does work on the tray when he applies a force that makes the tray move. This work transfers energy to the tray. The added energy is the kinetic energy of the moving tray.

Work done when you lift an object also increases the object's energy. Recall that the gravitational potential energy of an object increases as its height above the ground increases. When the girl in **Figure 4** lifts the tray the gravitational potential energy of the tray increases. The tray also gains kinetic energy as the tray moves upward. The girl did work on the tray as she lifted it. She transferred energy to the tray, increasing the tray's potential and kinetic energy.

🔑 **Key Concept Check** How does doing work on an object change its energy?

**Figure 4** Doing work on a tray transfers energy to it. The energy change can be in the form of kinetic energy or potential energy.

✓ **Visual Check** How does the tray's energy change when it is lifted?

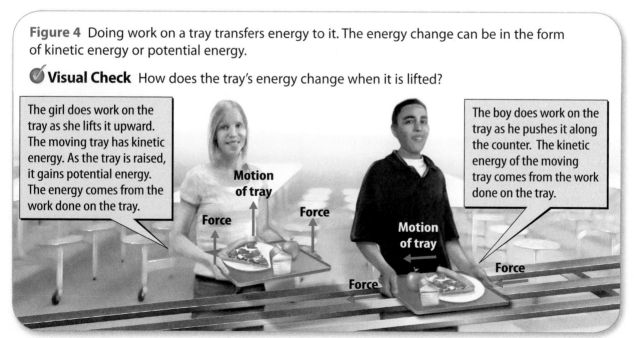

The girl does work on the tray as she lifts it upward. The moving tray has kinetic energy. As the tray is raised, it gains potential energy. The energy comes from the work done on the tray.

The boy does work on the tray as he pushes it along the counter. The kinetic energy of the moving tray comes from the work done on the tray.

Motion of tray

Force

Force

Motion of tray

Force

Force

# What is power?

What does it mean to have power? If two weight lifters lift identical weights to the same height, they both do the same amount of work. How quickly a weight lifter lifts a weight does not change the amount of work done. Doing work more quickly, does affect **power**—*the rate at which work is done.* A weight lifter lifting a weight more quickly exerts more power.

Knowing the power required for a task enables engineers to properly size engines and motors. You can calculate power by dividing the work done by the time needed to do the work. In the power equation, work done is in joules (J) and time is in seconds (s). The SI unit of power is the watt (W). Note that 1 J/s = 1 W.

**SCIENCE USE V. COMMON USE**

power
*Science Use* the rate at which work is done

*Common Use* the ability to accomplish something or to command or control other people

## Power Equation

power (in watts) = $\dfrac{\text{work (in joules)}}{\text{time (in seconds)}}$

$P = \dfrac{W}{t}$

You know that power is the rate at which work is done. You also know that work **transfers** energy. Thus, you can think of power as the rate at which energy is transferred to an object.

**ACADEMIC VOCABULARY**

transfer
*(verb)* to move from one place to another

 **Key Concept Check** How are work and power related?

---

**Math Skills** ✖️➗➕ **Power Equation**

**Solve for Power** A boy does 18 J of work in 2.0 s on his backpack as he lifts it from a table. How much power did the boy use on the backpack?

1 **This is what you know:**     work:    $W = 18$ J     time:    $t = 2.0$ s

2 **This is what you need to find:**    power:    $P$

3 **Use this formula:**     $P = \dfrac{W}{t}$

4 **Substitute:**     $P = \dfrac{18 \text{ J}}{2.0 \text{ s}} = 9.0$ J/s

   values for *W* and *t* into the formula and divide

> The symbol for work, *W*, is usually italicized. However, the abbreviation for watt, W, is not italicized.

5 **Convert units:**     J/s = W     $\dfrac{18 \text{ J}}{2.0 \text{ s}}$

**Answer:** The amount of power used was 9.0 W.

 Review

- Math Practice
- Personal Tutor

## Practice

A child pulls a wagon, doing 360 J of work in 8.0 s. How much power is exerted?

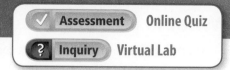

## Visual Summary

Work is done on an object when the object moves in the direction of the applied force.

When work is done on an object, energy is transferred to the object.

To increase power, work must be done in less time.

**FOLDABLES**

Use your lesson Foldable to review the lesson. Save your Foldable for the project at the end of the chapter.

## What do you think NOW?

You first read the statements below at the beginning of the chapter.

**1.** Work is done when you push a book across a table.

**2.** Doing work faster requires more power.

Did you change your mind about whether you agree or disagree with the statements? Rewrite any false statements to make them true.

## Use Vocabulary

**1** **Use the term** *work* in a sentence.

**2** **Define** *power* in your own words.

## Understand Key Concepts

**3** **Explain** how work and power are related.

**4** Lifting a stone block 146 m to the top of the Great Pyramid required 146,000 J of work. How much work was done to lift the block halfway to the top?
- **A.** 36,500 J
- **C.** 146,000 J
- **B.** 73,000 J
- **D.** 292,000 J

**5** **Give** an example of a situation in which doing work on an object changed its energy. Explain how the energy changed.

## Interpret Graphics

**6** **Explain** why this motionless weight lifter is not doing work.

**7** **Determine Cause and Effect** Copy and complete the graphic organizer below to list two ways power can be increased.

| Increase power |

## Critical Thinking

**8** **Explain** When you type on a computer keyboard, do you do work on the computer keys? Explain your answer.

## Math Skills

**Review**
—Math Practice—

**9** A motor applies a 5000-N force and raises an elevator 10 stories. If each story is 4 m tall, how much work does the motor do?

**10** Calculate the power, in watts, needed to mow a lawn in 50 minutes if the work required is 500,000 J.

# What is horsepower?

*You might be surprised to learn that there is a connection between a horse and a steam engine.*

In the early 1700s, horses did work on farms, powered factories, and moved vehicles. When people spoke of power, they literally were referring to horsepower. It was natural, then, for James Watt to think in terms of horsepower when he set out to improve the steam engine.

Watt did not invent the steam engine, but he realized its potential. He also realized that he needed a way to measure the power produced by steam engines. Watt knew that fabric mills used horses to power machinery. A worker attached a horse to a power wheel. The horse turned the wheel by walking in a 24-ft diameter circle at a rate of two revolutions per minute. From this information, Watt calculated the power supplied by a horse to be about 33,000 foot-pounds per minute. This amount of power became known as 1 horsepower (1 hp).

Watt succeeded in making better steam engines. Eventually, some steam engines produced more than 200 hp. Something unexpected happened as a result of all this power—life changed. More work was done and done faster. The mills expanded. People moved to the cities to work in the mills. Populations of cities in industrialized countries increased. The world changed because steam engines easily could produce more horsepower than horses.

▼ **A draft horse powering a mill**

▲ **An early steam engine developed by James Watt**

▼ **With cheap power came factories, jobs, and pollution.**

## It's Your Turn

**RESEARCH** Steam engines are seldom used today. Research steam engines to determine why the power source that was key to the Industrial Revolution is no longer used.

# Using Machines

## Reading Guide

### Key Concepts
ESSENTIAL QUESTIONS

- What are three ways a machine can make doing work easier?

- What is mechanical advantage?

- Why can't the work done by a machine be greater than the work done on the machine?

### Vocabulary

**mechanical advantage** p. 98

**efficiency** p. 99

g **Multilingual eGlossary**

## Inquiry **Could it be worse?**

Have you ever shoveled snow? You probably thought it was hard work. But imagine moving all of the snow only using your hands instead of a shovel. A shovel is a simple machine that makes moving snow easier.

### How do machines work?

Bicycles, pencil sharpeners, staplers, and doorknobs are machines that make specific tasks easier to do. How does the way a machine operates make work easier to do?

1 Read and complete a lab safety form.

2 Examine a **can opener** and answer these questions.

- How many different ways does the can opener move?
- Where do you apply force to the can opener?
- Where does the can opener apply force to the can?
- Is the amount of force you apply to the can opener different from the force the can opener applies to the can?
- Does the can opener change the direction of the forces applied to it?

3 Use the can opener to open a **can.** Record the process step by step in your Science Journal.

**Think About This**

1. Why is a can opener considered a machine? What task does it make easier to do?

2. 🔑 **Key Concept** Review your observations. Identify several ways the can opener makes opening a can easier.

## What is a machine?

If you ride a bicycle to school, you have firsthand experience with a machine. A machine is any device that makes doing something easier. The snow shovel on the previous page is a machine that is used to move snow. The scissors and the watch shown in **Figure 5** are also machines. As these example show, some machines are simple and other machines are more complex.

Like a pair of scissors, a leaf rake, broom, screwdriver, baseball bat, shovel, and door-knob are all machines. Other machines, such as a watch, an automobile, a snowblower, and a lawn mower, are more complex. All machines make tasks easier, but they do not decrease the amount of work required. Instead, a machine changes the way in which the work is done.

✓ **Reading Check** Do machines decrease the amount of work needed for a task? Explain.

**Figure 5** A machine makes work easier to do regardless of whether it is simple or complex.

### Does a ramp make it easier to lift a load?

In this lab, you will use a ramp and determine why it is useful for lifting heavy loads.

1. Read and complete a lab safety form.

2. Attach a **toy car** to the hook of a **spring scale.** Slowly lift the car off the table. Record in your Science Journal the force shown on the scale.

3. Lean a 30–40-cm **wood board** against a stack of three **textbooks** as shown. Use the scale to slowly pull the car up the ramp at a constant speed. Record the force.

### Analyze and Conclude

1. **Compare** How did the force needed to lift the car off the table compare with the force needed to pull the car up the ramp?

2. 🔑 **Key Concept** How does a ramp make work easier? Does the ramp decrease the amount of work done? Explain.

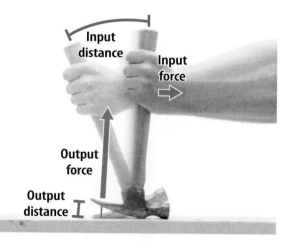

**Figure 6** You exert an input force on the hammer. The hammer exerts an output force on the nail.

### Input Force to Output Force

To use a machine, such as a hammer, you must apply a force to it. This force is the input force. The machine changes the input force into an output force, as shown in **Figure 6.** You apply an input force when you pull on the hammer's handle. The hammer changes the input force to an output force that pulls the nail out of the board.

### Input Work to Output Work

Squeezing the handles of a pair of scissors makes the blades move. You apply an input force that moves part of the machine—the scissors—and does work. The work is called input work, $W_{in}$. It is the product of the input force and the distance the machine moves in the direction of the input force.

Machines convert, or change, input work to output work. They do this by applying an output force on something and making it move. The output work, $W_{out}$, is the product of the output force and the distance part of the machine moves in the direction of the output force. The examples in **Figure 7** on the next page show these relationships.

## How do machines make work easier to do?

A machine can make work easier in three ways. It can change

- the size of a force;
- the distance the force acts;
- the direction of a force.

How hard it would be to pull a nail out of a board using your fingers? A hammer makes it easier in three ways. It changes the sizes of the forces and the distances the forces act. Notice that the person applies a smaller force over a longer distance, and the hammer exerts a greater force over a shorter distance. The hammer changes the direction of the input force. You pull back on the handle, and the hammer pulls up on the nail.

# 1 Change the Size of a Force

Could you use only your hands to pull a nailed-down board from a deck? Probably not. However, a crowbar makes this task fairly easy, as shown in **Figure 7**. You would start by placing the tip of a crowbar under the edge of the board. Then you would press down on the opposite end of the crowbar. The tip of the crowbar would lift the board away from the supporting board below. Repeating this process, you could remove the length of the board from the deck.

A crowbar is a machine. It makes work easier by changing a input force into a larger output force. Note that although the output force is greater than the input force, it acts over a shorter distance.

**Reading Check** How does a crowbar make work easier?

# 2 Change the Distance a Force Acts

Using a rake to gather leaves is an example of a machine that increases the distance over which a force acts. As shown in **Figure 7**, a person's hands move one end of the rake a short distance. The other end of the rake, however, sweeps through a greater distance making it easier to rake leaves. Note that the force applied by the rake (the output force) decreases as the distance over which the force acts (the output distance) increases. This relationship is true for all machines.

# 3 Change the Direction of a Force

Machines also can make work easier by changing the direction of the input force. A machine is used in **Figure 7** to lift a load. A rope tied to an object passes through a pulley. As the free end of the rope is pulled down, the object tied to the other end of the rope is lifted up. The machine changes the direction of the applied force.

**Key Concept Check** In what three ways do machines make doing work easier?

**Figure 7** Machines make work easier in three ways.

**1** When the output force is greater than the input force, the output force acts over a shorter distance.

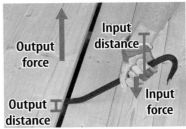

Input force × Input distance = Output force × Output distance

Input work = Output work

**2** When the output force acts over a longer distance than the input force, the output force is less than the input force.

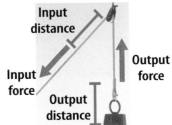

Input force × Input distance = Output force × Output distance

Input work = Output work

**3** Equal output and input forces act over equal distances.

Input force × Input distance = Output force × Output distance

Input work = Output work

# What is mechanical advantage?

**WORD ORIGIN**

mechanical
from Greek *mechanikos*, means
"machine"

Most machines change the size of the force applied to them. *A machine's* **mechanical advantage** *is the ratio of a machine's output force produced to the input force applied.* The mechanical advantage, or *MA*, tells you how many times larger or smaller the output force is than the input force.

**FOLDABLES**

Make a three-tab book from a sheet of horizontal paper. Label it as shown. Use it to summarize the three ways machines make doing work easier.

How Machines Make Work Easier

| Changing Size of Force | Changing Distance | Changing Direction |

---

### Mechanical Advantage Equation

mechanical advantage (no units) = $\dfrac{\text{output force (in newtons)}}{\text{input force (in newtons)}}$

$$MA = \frac{F_{out}}{F_{in}}$$

---

Mechanical advantage can be less than 1, equal to 1, or greater than 1. A mechanical advantage greater than 1 means the output force is greater than the input force. A crowbar, for example, has a mechanical advantage much greater than 1.

The ideal mechanical advantage, or *IMA*, is the mechanical advantage if no friction existed. Machines cannot operate at ideal mechanical advantage because friction always exists.

🔑 **Key Concept Check** What is mechanical advantage?

---

**Math Skills** ➗  Mechanical Advantage Equation

**Solve for Mechanical Advantage** A carpenter applies 525 N to the end of a crowbar. The force exerted on the board is 1,575 N. What is the mechanical advantage of the crowbar?

**1** **This is what you know:**
input force:   $F_{in} = 525$ N
output force:   $F_{out} = 1{,}575$ N

**2** **This is what you need to find:**   mechanical advantage:   *MA*

**3** **Use this formula:**   $MA = \dfrac{F_{out}}{F_{in}}$

**4** **Substitute:**   $MA = \dfrac{1{,}575 \text{ N}}{525 \text{ N}} = 3$
the values for $F_{in}$ and $F_{out}$
into the formula and divide

**Answer:** The mechanical advantage is 3.

**Review**
• Math Practice
• Personal Tutor

## Practice

While raking leaves, a woman applies an input force of 32 N to a rake. The rake has an output force of 16 N. What is the mechanical advantage of the rake?

# What is efficiency?

The output work done by a machine never exceeds the input work of the machine. The reason for this is friction. Friction converts some of the input work to thermal energy. The converted energy cannot be used to do work.

The **efficiency** *of a machine is the ratio of the output work to the input work.* Efficiency is calculated using the equation below. Because output work is always less than input work, a machine's efficiency is always less than 100 percent. As shown in **Figure 8**, lubricating a machine's moving parts increases efficiency.

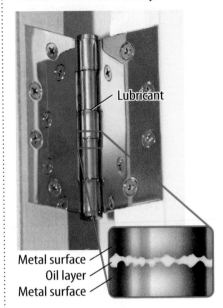

Lubricant

Metal surface
Oil layer
Metal surface

## Efficiency Equation

$$\text{efficiency (in \%)} = \frac{\text{output work (in joules)}}{\text{input work (in joules)}} \times 100\%$$

$$\text{efficiency} = \frac{W_{out}}{W_{in}} \times 100\%$$

 **Key Concept Check** Why can't the work done by a machine be greater than the work done on the machine?

---

## Math Skills　　Efficiency Equation

**Solve for Efficiency** A mechanic does 78.0 J of work pulling the rope on a pulley to lift a motor. The output work of the pulley is 64.0 J. What is the efficiency of the pulley?

**1** **This is what you know:**　　input work:　$W_{in} = 78.0$ J

output work:　$W_{out} = 64.0$ J

**2** **This is what you need to find:**　efficiency

**3** **Use this formula:**　　$\text{efficiency} = \dfrac{W_{out}}{W_{in}} \times 100\%$

**4** **Substitute:**　　$\text{efficiency} = \dfrac{64.0 \text{ J}}{78.0 \text{ J}} \times 100\% = 82.1\%$
the values for $W_{in}$ and $W_{out}$
into the formula and divide

**Answer:** The efficiency is 82.1%

 Review
- Math Practice
- Personal Tutor

### Practice
A carpenter turns a handle to adjust a saw blade. The input work is 55 J and the output work is 51 J. What is the efficiency of the blade adjuster?

# Lesson 2 Review

## Visual Summary

A machine can be simple or complex. It makes a task easier.

The mechanical advantage of a machine indicates how it changes an input force.

Lubricant

The efficiency of a machine is increased when a lubricant coats moving parts.

**FOLDABLES**

Use your lesson Foldable to review the lesson. Save your Foldable for the project at the end of the chapter.

## What do you think NOW?

You first read the statements below at the beginning of the chapter.

**3.** Machines always decrease the force needed to do a job.

**4.** A well-oiled, low-friction machine can be 100 percent efficient.

Did you change your mind about whether you agree or disagree with the statements? Rewrite any false statements to make them true.

## Use Vocabulary

1. **Define** *efficiency* in your own words.

2. **Distinguish** between efficiency and mechanical advantage.

## Understand Key Concepts

3. **Explain** how friction reduces the efficiency of machines.

4. Which machine efficiency is impossible?
   A. 1 percent       C. 99 percent
   B. 80 percent      D. 100 percent

5. **Compare and contrast** input and output forces and input and output distances for a hammer pulling a nail out of a board.

## Interpret Graphics

6. **Analyze** How does a rake make gathering leaves easier? Explain in terms of distances and forces.

7. **Organize** Copy and fill in the graphic organizer below to describe three ways machines can make work easier.

Making work easier

## Critical Thinking

8. **Modify** the efficiency equation by writing it in terms of input force and output force.

## Math Skills ✕ ÷

 Review
— Math Practice —

9. A pulley system uses a 250-N force to lift a 2,750-N crate. What is the mechanical advantage of the system?

10. An assembly line machine needs 150 J of input work to do 90 J of output work. What is the efficiency of the machine?

# How does mechanical advantage affect a machine?

### Materials

centimeter ruler

clear plastic tape

modeling clay

spring scale

### Safety

Machines change forces and make work easier. In this lab, you build and use a simple machine to lift a load, and record force and distance data. You then interpret data and determine how the mechanical advantage of the machine affected the results.

## Learn It

Measurements and observations are data. When you **interpret data,** you look for patterns and relationships in the data.

## Try It

1 Read and complete a lab safety form.

2 Lay a metric ruler on a table so one end extends over an edge, as shown below. Tape the other end of the ruler to the table. Make sure the free end of the ruler moves up and down easily. This is your simple machine.

3 Use a spring scale to measure the mass of a handful-sized lump of clay. Use this weight as the output force ($F_{out}$) for your tests.

4 Push the clay onto the 15-cm mark of the ruler so that it sticks.

5 Attach a spring scale to the ruler's free end. Pull upward on the spring scale so the ruler lifts off of the table. Record the force reading as the input force ($F_{in}$).

6 Make a data table in your Science Journal. Record your measurements, including the distance from the taped end to the clay ($d_{out}$) and from the taped end to the spring scale ($d_{in}$).

7 Calculate the mechanical advantage ($MA$) of the machine from your data. $MA = \dfrac{F_{out}}{F_{in}}$.

## Apply It

8 Move the clay to the 27-cm mark. Repeat steps 5 through 7.

9 Move the clay to two other locations on the ruler, repeating steps 5 through 7 each time.

10 How is the machine's mechanical advantage related to $d_{out}$, the distance from the taped end of the ruler to the clay?

11 When the spring scale lifts the end of the ruler up, is the clay lifted the same amount? Does the amount the clay is lifted depend on its location along the ruler? Explain.

12 🔑 **Key Concept** What is the relationship between the machine's mechanical advantage and how easy it is to lift the clay?

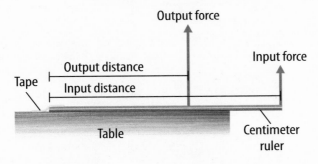

Output force

Input force

Output distance

Input distance

Tape

Table

Centimeter ruler

# Simple Machines

## Reading Guide

### Key Concepts 🔑
**ESSENTIAL QUESTIONS**

- What is a simple machine?
- How is the ideal mechanical advantage of simple machines calculated?
- How are simple machines and compound machines different?

### Vocabulary

**simple machine** p. 103

**lever** p. 104

**fulcrum** p. 104

**wheel and axle** p. 106

**inclined plane** p. 107

**wedge** p. 108

**screw** p. 108

**pulley** p. 109

 **Multilingual eGlossary**

 **Video**

- BrainPOP®
- Science Video

### Inquiry Hit or Miss?

A trebuchet (TRE bu shet) is a device from the Middle Ages designed to hurl large rocks. The trebuchet was effective at hurling rocks at castle walls, though it had limited accuracy. A trebuchet is a large simple machine.

## How does a lever work?

Humans have used levers for thousands of years to make work easier. How does a lever change the force applied to it?

1. Read and complete a lab safety form.

2. Place a **ruler** on top of an **eraser** as shown. Place a **book** on one end of the ruler.

3. Lift the book by pushing down on the other end of the ruler. Record in your Science Journal how easy or difficult it is to lift the book.

4. Repeat steps 2 and 3 several times using different locations of the eraser along the ruler.

**Think About This**

1. Explain how changing the position of the eraser affected the force exerted by the lever.

2.  **Key Concept** How would you describe the motion of the lever? Is the motion of the lever simple or complicated? Explain.

## What is a simple machine?

Some machines, such as a trebuchet, have only a few parts. Six types of **simple machines**, shown in **Figure 9,** *do work using only one movement.* They are lever, wheel and axle, inclined plane, wedge, screw, and pulley.

 **Key Concept Check** Describe a simple machine.

**Figure 9**  The six types of simple machines are shown below. Each simple machine does work with one motion.

**Lever**

**Wheel and axle**

**Inclined plane**

**Wedge**

**Screw**

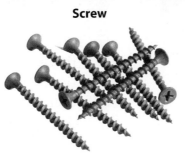

**Pulley**

## The Three Classes of Levers

**Figure 10** The location of the input force, the output force, and the fulcrum determine the class of lever.

**1** **First-Class Lever**

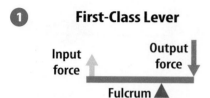

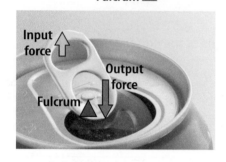

**2** **Second-Class Lever**

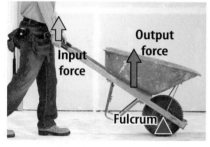

**3** **Third-Class Lever**

## Levers

The next time you open an aluminum beverage can, watch how the finger tab works. The finger tab is a **lever**—*a simple machine made up of a bar that pivots, or rotates, about a fixed point. The point about which a lever pivots is called a* **fulcrum.** The fulcrum on the aluminum beverage can in **Figure 10** is where the finger tab attaches to the can.

Notice that the input force and the output force act on opposite ends of the finger tab. The distance from the fulcrum to the input force on the tab is the *input arm*. The distance from the fulcrum to the end of the tab that pushes down on the can is the *output arm*.

There are three types of levers, also shown in **Figure 10.** First-class, second-class, and third-class levers differ in where the input force and output force are relative to the fulcrum.

**1** In a first-class lever, the fulcrum is between the input force and the output force. The direction of the input force is always different than the direction of the output force. A hammer is a first-class lever when it is used to pull a nail out of wood. A finger tab on a beverage can is also a first-class lever.

**2** A second-class lever has the output force between the input force and the fulcrum. The output force and the input force act in the same direction. A second-class lever makes the output force greater than the input force. A wheelbarrow, a nut cracker, and your foot are second-class levers.

**3** The input force is between the output force and the fulcrum in a third-class lever. The output force is less than the input force, though both forces act in the same direction. Tweezers, a rake, and a broom are examples of third-class levers.

✓ **Reading Check** What is a lever?

| Mechanical Advantage of a First-Class Lever | Mechanical Advantage of a Second-Class Lever | Mechanical Advantage of a Third-Class Lever |
|---|---|---|

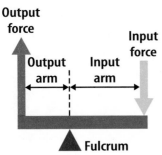

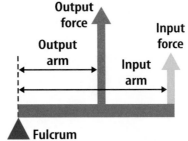

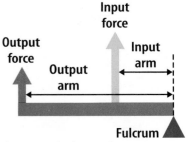

- Fulcrum location determines mechanical advantage.
- Mechanical advantage can be less than 1, equal to 1, or greater than 1.

- Input arm is longer than output arm.
- Output force is greater than input force.
- Mechanical advantage is greater than 1.

- Input arm is shorter than output arm.
- Output force is less than input force.
- Mechanical advantage is less than 1.

## Mechanical Advantage of Levers

The ideal mechanical advantage of a lever equals the length of the input arm divided by the length of the output arm.

### Ideal Mechanical Advantage of a Lever

ideal mechanical advantage = $\dfrac{\text{length of input arm (in meters)}}{\text{length of output arm (in meters)}}$

$$IMA = \frac{L_{in}}{L_{out}}$$

**First-Class Levers** The location of the fulcrum determines the mechanical advantage in first-class levers. In **Figure 11,** the mechanical advantage of the lever is greater than 1 because the input arm is longer than the output arm. This makes the output force greater than the input force. When the mechanical advantage is equal to 1, the input and output arms and input and output forces are equal. If the mechanical advantage is less than 1, the input arm is shorter than the output arm. The output force then is less than the input force.

**Second-Class Levers** The input arm is longer than the output arm for all second-class levers. The mechanical advantage of a second-class lever is always greater than 1.

**Third-Class Levers** For third-class levers, the input arm is always shorter than the output arm. Thus, the mechanical advantage of a third-class lever is always less than 1.

 **Key Concept Check** How is the ideal mechanical advantage of a lever calculated?

**Figure 11** The mechanical advantage of a lever varies depending on the location of the fulcrum.

✔ **Visual Check** How does the input arm compare to the output arm in a second-class lever? In a third-class lever?

🔲 **Review**

**Personal Tutor**

**First-Class Lever**

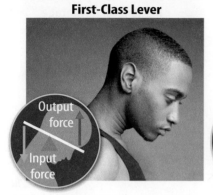

**Second-Class Lever**

**Third-Class Lever**

▲ **Figure 12** The neck, foot, and arm are examples of first-, second-, and third-class levers in the human body.

## Levers in the Human Body

The human body uses all three classes of levers to move. Muscles provide the input force for the levers. Examples of levers in the neck, foot, and arm are shown in **Figure 12.**

**The Neck** Your neck acts like a first-class lever. The fulcrum is the joint connecting the skull to the spine. The neck muscles provide the input force. The output force is applied to the head and helps support your head's weight.

**The Foot** When standing on your toes, the foot acts like a second-class lever. The ball of the foot is the fulcrum. The input force comes from muscles on the back of the lower leg.

**The Arm** Your forearm works like a third-class lever. The elbow is the fulcrum and the input force comes from muscles located near the elbow.

**Figure 13** A screwdriver is a wheel and axle. The handle is the wheel and the shaft is the axle. ▼

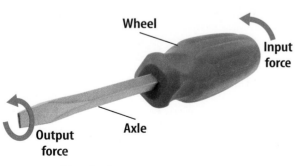

## Wheel and Axle

*A* **wheel and axle** *is an axle attached to the center of a wheel and both rotate together.* Note that *axle* is another word for a shaft. The screwdriver shown in **Figure 13** is a wheel and axle. The handle is the wheel because it has the larger diameter. The axle is the shaft attached to the handle. Both the handle and the shaft rotate when the handle turns.

### Mechanical Advantage of a Wheel and Axle

For a wheel and axle, the length of the input arm is the radius of the wheel. Likewise, the length of the output arm is the radius of the axle. These lengths give the ideal mechanical advantage as shown in the equation below.

---

**Ideal Mechanical Advantage of Wheel and Axle**

$$\text{ideal mechanical advantage} = \frac{\text{radius of wheel (in meters)}}{\text{radius of axle (in meters)}}$$

$$IMA = \frac{r_{wheel}}{r_{axle}}$$

---

### Using a Wheel and Axle

You know a screwdriver makes it easier to turn a screw. But how does a screwdriver work? Examining the mechanical advantage equation offers an explanation. When you turn a screwdriver, you apply an input force to the handle (the wheel). The output force is applied to the screw by the screwdriver's shaft (the axle). Because the wheel is larger than the axle, the mechanical advantage is greater than 1. This makes the screw easier to turn.

## Inclined Planes

The ancient Egyptians built pyramids using huge stone blocks. Moving the blocks up the pyramid must have been difficult. To make the task easier, ramps were often used. *A ramp, or* **inclined plane,** *is a flat, sloped surface.* It takes less force to move an object upward along an inclined plane than it does to lift the object straight up. As shown in **Figure 14,** ramps are still useful for moving heavy loads.

**Figure 14** Moving a sofa is easier using a ramp. As shown, using a ramp only requires a 100-N force to move the 500-N sofa. Because of friction, no ramp operates at its ideal mechanical advantage.

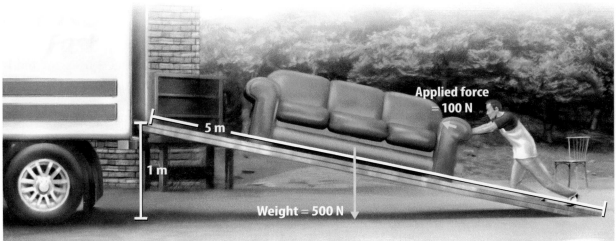

Applied force = 100 N

5 m

1 m

Weight = 500 N

Input force

## Mechanical Advantage of Inclined Planes

The ideal mechanical advantage of an inclined plane equals its length divided by its height. See the equation below.

### Ideal Mechanical Advantage of an Inclined Plane

$$\text{ideal mechanical advantage} = \frac{\text{length of inclined plane (in meters)}}{\text{height of inclined plane (in meters)}}$$

$$IMA = \frac{\ell}{h}$$

Note that increasing the length and decreasing the height of the inclined plane increases its ideal mechanical advantage. The longer or less-sloped an inclined plane is, the less force is needed to move an object along its surface.

## Wedges

*A sloped surface that moves is called a* **wedge.** A wedge is really a type of inclined plane with one or two sloping sides. A doorstop is a wedge with one sloped side. The wedge shown in **Figure 15** is a wedge with two sloped sides. Notice how the shape of the wedge gives the output forces a different direction than the input force.

Your front teeth also are wedges. As you push your front teeth into food, the downward force is changed by your teeth into a sideways force that pushes the food apart.

✔ **Reading Check** How are a wedge and a ramp different?

## Screws

As shown in **Figure 16,** a **screw** *is an inclined plane wrapped around a cylinder.* When you turn a screw, the screw threads change the input force to an output force. The output force pulls the screw into the material.

Figure 16 The groove, or thread, that wraps around a screw is an inclined plane. ▼

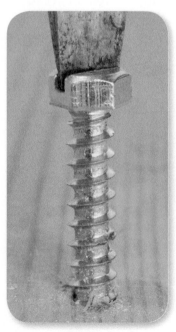

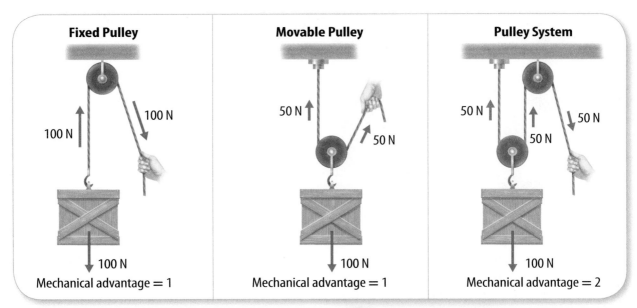

| Fixed Pulley | Movable Pulley | Pulley System |
|---|---|---|

100 N    100 N    100 N

Mechanical advantage = 1    Mechanical advantage = 1    Mechanical advantage = 2

**Figure 17** Pulleys can change force and direction.

## Pulleys

You might have seen large cranes lifting heavy loads at construction sites. The crane uses a **pulley**—*a simple machine that is a grooved wheel with a rope or a cable wrapped around it.*

**Fixed Pulleys** Have you ever pulled down on a cord to raise a window blind? The cord passes through a fixed pulley mounted to the top of the window frame. A fixed pulley only changes the direction of the force, as shown **Figure 17.**

**Movable Pulleys and Pulley Systems** A pulley can also be attached to the object being lifted. This type of pulley, called a movable pulley, is shown in **Figure 17.** Movable pulleys decrease the force needed to lift an object. The distance over which the force acts increases.

A pulley system is a combination of fixed and movable pulleys that work together. An example of a pulley system is shown in **Figure 17.**

**Mechanical Advantage of Pulleys** The ideal mechanical advantage of a pulley or a pulley system is equal to the number of sections of rope pulling up on the object.

**Inquiry MiniLab**
     **20 minutes**

### Can you increase mechanical advantage?

Pulley systems change the size and the direction of an applied force. How can you change the mechanical advantage of a pulley system?

1. Read and complete a lab safety form.

2. Use two **broomsticks** and an 8-m length of **rope** to make a pulley system as shown. Hold each broomstick waist-high. Tie the rope to one of the broomsticks and loop it around the other.

3. Have a student pull the free end of the rope as shown. Record in your Science Journal the forces acting on the broomsticks.

4. Take the excess rope and make two more loops around the broomsticks. Repeat step 3.

#### Analyze and Conclude

1. **Describe** how the forces applied on the broomsticks changed when the number of loops in the pulley system increased.

2. 🔑 **Key Concept** Relate the number of rope segments in each pulley system to the mechanical advantage of each system.

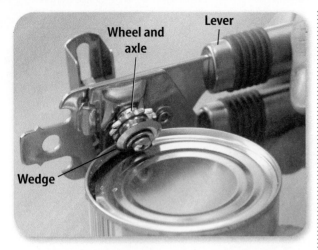

▲ **Figure 18** A can opener is a compound machine. It uses a second-class lever to move the handle, a wheel and axle to turn the blade, and a wedge to puncture the lid.

🔍 **Visual Check** How many simple machines are part of the can opener?

**Figure 19** A system of gears is a compound machine. Notice how the direction of rotation changes from one gear to the next. ▼

# What is a compound machine?

Two or more simple machines that operate together form a compound machine. The can opener in **Figure 18** is a compound machine.

🔑 **Key Concept Check** How are simple machines and compound machines different?

## Gears

A gear is a wheel and axle with teeth around the wheel. Two or more gears working together form a compound machine. When the teeth of two gears interlock, turning one gear causes the other to turn. The direction of motion, as shown in **Figure 19,** changes from one gear to the next.

Gears of different sizes turn at different speeds. Smaller gears rotate faster than larger gears. The amount of force transmitted through a set of gears is also affected by the size of the gears. The input force applied to a large gear is reduced when it is applied to a smaller gear.

## Efficiency of Compound Machines

How can you determine the efficiency of a compound machine? The efficiency of a compound machine is calculated by multiplying the efficiencies of each simple machine together.

Consider the can opener in **Figure 18.** It is made up of three simple machines. Suppose the efficiencies are as follows:

- efficiency of lever = 95%
- efficiency of wheel and axle = 90%
- efficiency of wedge = 80%

The efficiency of the can opener is equal to the product of these efficiencies.

$$\text{efficiency of the can opener} = 95\% \times 90\% \times 80\% = 68\%$$

Each simple machine decreases the overall efficiency of the compound machine.

# Lesson 3 Review

## Visual Summary

Six simple machines are the lever, wheel and axle, inclined plane, wedge, screw, and pulley.

There are three types of levers—first class, second class, and third class.

The kind of wedge used to split logs is a simple machine.

**FOLDABLES®**

Use your lesson Foldable to review the lesson. Save your Foldable for the project at the end of the chapter.

## What do you think NOW?

You first read the statements below at the beginning of the chapter.

**5.** A doorknob is a simple machine.

**6.** A loading ramp makes it easier to lift a load.

Did you change your mind about whether you agree or disagree with the statements? Rewrite any false statements to make them true.

## Use Vocabulary

**1** **Define** *wheel and axle* in your own words.

## Understand Key Concepts

**2** **Identify** the simple machine in which the fulcrum is between the input force and the output force.

   **A.** wheel and axle    **C.** inclined plane

   **B.** lever           **D.** pulley

**3** **Determine** how the ideal mechanical advantage of a ramp changes if the ramp is made longer.

**4** **Calculate** the ideal mechanical advantage of a screwdriver with a 24-mm radius handle and an 8-mm radius shaft.

**5** **Explain** how simple and compound machines are different.

## Interpret Graphics

**6** **Observe** the scissors shown. They have a pin about which the blades rotate. Each blade has a sloped cutting surface. Identify the simple machines in a pair of scissors.

**7** **Organize Information**
Copy and fill in the graphic organizer below with the equations used to calculate the ideal mechanical advantage of each these simple machines.

| Type of Simple Machine | Ideal Mechanical Advantage |
|---|---|
| Lever | |
| Wheel and axle | |
| Inclined plane | |

## Critical Thinking

**8** **Suggest** a way in which a pulley could be considered a type of lever.

## Materials

250-g
hanging mass

5-N spring
scale

small pulley

ring stand

meterstick

**Also needed:**
50-cm × 15-cm
board, books,
2-m length of
heavy string

**Safety**

# Comparing Two Simple Machines

You will use a pulley and an inclined plane to lift a 250-g mass to a height of 20 cm. Which simple machine makes the work of lifting the load the easiest?

## Question

Will the pulley or the inclined plane have a greater mechanical advantage?

## Procedure

1 Read and complete a lab safety form.

2 Examine the equipment and diagrams. With your group, discuss why each simple machine has a mechanical advantage when lifting the mass.

3 Predict whether the pulley or the inclined plane will provide the greatest mechanical advantage in lifting the mass. Record your prediction in your Science Journal.

4 Also in your Science Journal, make a data table with columns for data you plan to collect to calculate mechanical advantage for each simple machine.

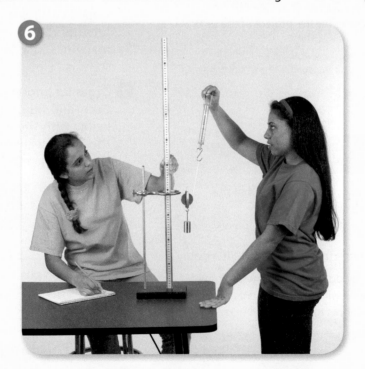

**5** Use a spring scale to measure the weight of the 250-g mass. Record the weight in newtons.

**6** Set up the pulley and the ring stand as shown on the previous page. Adjust the spring scale and the pulley so that the weight just clears the table. Use the spring scale as shown to slowly lift the weight to a height of 20 cm. Record the force in newtons. Repeat.

**7** Set up the inclined plane as shown to the right so that the top of the ramp is 20 cm above the table. Attach the spring scale to the weight. Beginning at the bottom of the ramp, use the spring scale to pull the weight slowly up the ramp. Record the force in newtons. Repeat.

## Analyze and Conclude

**8** **Interpret Data** What is the mechanical advantage of the pulley and the inclined plane? Was your prediction correct? Explain.

**9** **Explain** What other measurements would you need in order to calculate the efficiency of each machine?

**10** **The Big Idea** Did the amount of work done on the weight to lift it to a height of 20 cm change with each machine? Explain. Did the machines make doing the work easier?

## Communicate Your Results

Make a large data table on which all groups can display their data. Discuss similarities and differences between group data.

 **Extension**

How can you increase the mechanical advantage of your inclined plane? To investigate your question, design a controlled experiment.

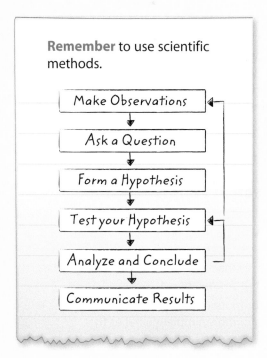

**Remember** to use scientific methods.

Make Observations

↓

Ask a Question

↓

Form a Hypothesis

↓

Test your Hypothesis

↓

Analyze and Conclude

↓

Communicate Results

# Chapter 3 Study Guide

**A machine makes work easier by changing the size of the applied force, changing the distance over which the applied force acts, or changing the direction of the applied force.**

## Key Concepts Summary 🔑

### Lesson 1: Work and Power

- For **work** to be done on an object, an applied force must move the object in the direction of the force.
- When work is done on an object, the energy of the object increases.
- **Power** is the rate at which work is done.

### Lesson 2: Using Machines

- A machine can make work easier in three ways: changing the size of a force, changing the distance the force acts, or changing the direction of a force.
- The **mechanical advantage** of a machine is the ratio of the output force to the input force.
- Because of friction, the output work done by a machine is always less than the input work to the machine.
- Friction between moving parts converts some of the input work into thermal energy and decreases the **efficiency** of the machine.

### Lesson 3: Simple Machines

- A **simple machine** does work using only one movement.
- The ideal mechanical advantage of simple machines is calculated using simple formulas.
- A compound machine is made up of two or more simple machines that operate together.

## Vocabulary

**work** p. 87
**power** p. 91

**mechanical advantage** p. 98
**efficiency** p. 99

**simple machine** p. 103
**lever** p. 104
**fulcrum** p. 104
**wheel and axle** p. 106
**inclined plane** p. 107
**wedge** p. 108
**screw** p. 108
**pulley** p. 109

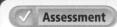
## FOLDABLES® Chapter Project

Assemble your lesson Foldables as shown to make a Chapter Project. Use the project to review what you have learned in this chapter.

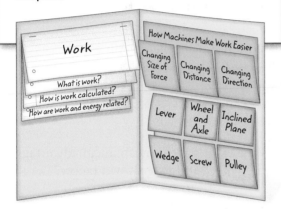

## Use Vocabulary

*Match each phrase with the correct vocabulary term from the Study Guide.*

**1** A wheel and axle is an example of a(n) _____.

**2** As work is done more quickly, the _____ required increases.

**3** A third-class lever has a(n) _____ that is always less than 1.

**4** The _____ of a machine is always less than 100 percent.

**5** A sloped road is an example of a(n) _____.

**6** Your front teeth act like a(n) _____.

**7** A(n) _____ is an inclined plane wrapped around a cylinder.

## Link Vocabulary and Key Concepts

((O)) **Concepts in Motion**    Interactive Concept Map

*Copy this concept map, and then use vocabulary terms from the previous page and other terms in this chapter to complete the concept map.*

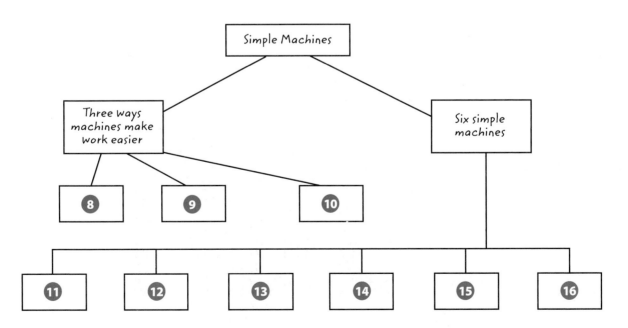

# Chapter 3 Review

## Understand Key Concepts 🔑

**1** Which must be true when work is done?
   A. No force acts on the object.
   B. An object slows down.
   C. An object moves quickly.
   D. An object moves in the same direction as a force exerted on it.

**2** Which must be true when work is done on a resting object and the object moves?
   A. The energy of the object changes.
   B. The energy of the object is constant.
   C. The force on the object is constant.
   D. The force on the object is zero.

**3** Which increases the power used in lifting the backpack?
   A. lifting at a constant speed
   B. lifting at an angle
   C. lifting more quickly
   D. lifting a lighter backpack

**4** Which does more work on an object?
   A. pushing a 50-N object a distance of 2 m
   B. pushing a 120-N object a distance of 7 m
   C. pushing a 150-N object a distance of 5 m
   D. pushing a 300-N object a distance of 1 m

**5** Which must be true if a machine's mechanical advantage is equal to 2?
   A. Direction of force is changed.
   B. Output force equals input force.
   C. Output force is greater than input force.
   D. Output work is greater than input work.

**6** Which is ideal mechanical advantage of the pulley system?
   A. 1
   B. 2
   C. 3
   D. 4

Force

**7** How does friction affect a machine?
   A. It converts input work into thermal energy.
   B. It converts output work into thermal energy.
   C. It converts thermal energy into input work.
   D. It converts thermal energy into output work.

**8** Which is a simple machine?
   A. tweezers
   B. bicycle
   C. can opener
   D. car

**9** What is the ideal mechanical advantage of the wheel and axle shown below?

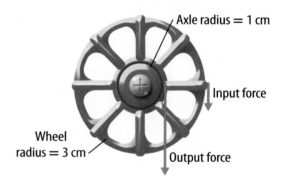
Axle radius = 1 cm
Input force
Wheel radius = 3 cm
Output force

   A. 1
   B. 2
   C. 3
   D. 4

**10** Which is most easily used as either a wheel and axle or as a lever?
   A. ax
   B. scissors
   C. screwdriver
   D. screw

**11** Which quantity increases when a frictionless ramp is used to lift an object?
   A. input force
   B. input distance
   C. input work
   D. output work

## Critical Thinking

**12** **Contrast** the work done when you lift a box with the work done when you carry the box across a room.

**13** **Analyze** Does a weight lifter transfer more energy, less energy, or the same amount of energy when lifting the weight faster? Explain.

**14** **Decide** First you lift an object from the floor onto a shelf. Then, you move the object from the shelf back to the floor. Do you perform the same amount of work each time? Explain.

**15** **Explain** why the output work done by a machine is never greater than the input work to the machine.

**16** **Suggest** a reason the efficiency of a machine used in a factory might decrease over time. What could be done to increase the efficiency?

**17** **Explain** Imagine using a screwdriver to drive a screw into a piece of wood. Explain why turning the handle of the screwdriver is easier than turning its shaft.

**18** **Explain** Can you determine if work is being done on the backpack simply by looking at this photo? Explain why or why not.

*Writing in Science*

**19** **Write** a paragraph about using an inclined plane on the Moon. Objects weigh less on the Moon because the force of gravity is less than on Earth. Explain whether this affects the mechanical advantage of the inclined plane or the way it is used.

**REVIEW** THE **BIG** IDEA

**20** How do machines make doing work easier? Describe several examples of how machines change forces and make work easier to do.

**21** Explain why the earthmover below uses more power than a person moving the same amount of dirt with a shovel.

**Math Skills**

Review
—— Math Practice ——

### Use Numbers

**22** How much work is done when a force of 30 N moves an object a distance of 3 m?

**23** How much power is used when 600 J of work is done in 10 s?

**24** How much force would be needed to push a box weighing 30 N up a ramp that has a ideal mechanical advantage of 3? Assume there is no friction.

**25** Calculate the efficiency of a machine that requires 20 J of input work to do 10 J of output work.

**26** Calculate the mechanical advantage of a machine that changes an input force of 50 N into an output force of 150 N.

**27** A compound machine is made up of four simple machines. If the efficiencies of the simple machines are 98%, 93%, 87%, and 92%, respectively, what is the overall efficiency of the compound machine?

# Standardized Test Practice

*Record your answers on the answer sheet provided by your teacher or on a sheet of paper.*

## Multiple Choice

**1** Which is the SI unit of work?

  **A** ampere

  **B** joule

  **C** newton

  **D** watt

**2** Which transfers both kinetic energy and potential energy to an object?

  **A** lifting it

  **B** lowering it

  **C** pushing it

  **D** rolling it

*Use the chart below to answer questions 3 and 4.*

| Moving a Chair | |
|---|---|
| Force | 20 newtons |
| Distance | 5 meters |
| Time | 2 seconds |

**3** How much work was involved in moving the chair?

  **A** 4 J

  **B** 10 J

  **C** 40 J

  **D** 100 J

**4** How much power was used to move the chair?

  **A** 10 W

  **B** 20 W

  **C** 50 W

  **D** 200 W

**5** Which is a third-class lever?

  **A** a broom

  **B** a hammer

  **C** a nutcracker

  **D** a wheelbarrow

*Use the chart below to answer question 6.*

| Mechanical Advantage Equation |
|---|
| mechanical advantage $= \dfrac{\text{output force}}{?}$ |

**6** Which correctly completes the mechanical advantage equation?

  **A** distance

  **B** time

  **C** input force

  **D** output work

**7** A simple machine is NOT able to

  **A** change the size of a force.

  **B** decrease the amount of work required.

  **C** exert an output force on an object.

  **D** make work easier to perform.

**8** Which MUST happen in order for work to be done?

  **A** A force must move an object.

  **B** A machine must transfer force.

  **C** Force must be applied to an object.

  **D** Output force must exceed input force.

**9** Which increases the efficiency of a complex machine?

  **A** adding more simple machines

  **B** increasing input force

  **C** putting more work in

  **D** reducing friction by lubricating

*Use the diagram below to answer question 10.*

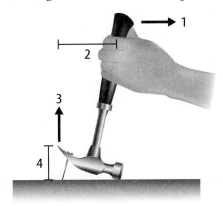

**Constructed Response**

*Use the table below to answer questions 13 and 14.*

| Simple Machine | Example | Task |
|---|---|---|
| Lever | | |
| Inclined plane | | |
| Wheel and axle | | |
| Pulley | | |
| Wedge | | |
| Screw | | |

**10** Which number represents the output distance in the diagram above?

   **A** 1

   **B** 2

   **C** 3

   **D** 4

**11** What information is needed to calculate the ideal mechanical advantage of an inclined plane?

   **A** its height and length

   **B** its length and thickness

   **C** its thickness and weight

   **D** its weight and width

**12** Which is a characteristic feature of a wedge?

   **A** cable

   **B** fulcrum

   **C** slope

   **D** thread

**13** In the table, the six types of simple machines are listed. Provide an example of each machine. Then, list an everyday task you have performed with the help of the machine.

**14** How do each of the simple machines in the table above make work easier?

**15** What is the difference between simple and compound machines?

**16** Describe a real-life situation in which simple machines would make work easier. Identify two machines that would be helpful in the situation and explain how they would be used.

**17** What two factors affect the amount of work done on an object? When does a force acting on an object NOT do work? Give an example.

| NEED EXTRA HELP? | | | | | | | | | | | | | | | | | |
|---|---|---|---|---|---|---|---|---|---|---|---|---|---|---|---|---|---|
| If You Missed Question... | 1 | 2 | 3 | 4 | 5 | 6 | 7 | 8 | 9 | 10 | 11 | 12 | 13 | 14 | 15 | 16 | 17 |
| Go to Lesson... | 1 | 1 | 1 | 1 | 3 | 2 | 2 | 1 | 2 | 2 | 3 | 3 | 3 | 3 | 3 | 2,3 | 1 |

# Chapter 4

# Forces and Fluids

**THE BIG IDEA** In what ways do people use forces in fluids?

## Inquiry What makes it float?

Did you know that the helium in parade balloons is considered a fluid? Forces acting on the helium in this balloon keep it in the air. What are those forces?

- What is a fluid?

- How do forces affect fluids and the objects within them?

- In what ways do people use forces in fluids?

## Get Ready to Read

### What do you think?

Before you read, decide if you agree or disagree with each of these statements. As you read this chapter, see if you change your mind about any of the statements.

**1** Air is a fluid.

**2** Pressure is a force acting on a fluid.

**3** You can lift a rock easily under water because there is a buoyant force on the rock.

**4** The buoyant force on an object depends on the object's weight.

**5** If you squeeze an unopened plastic ketchup bottle, the pressure on the ketchup changes everywhere in the bottle.

**6** Running with an open parachute decreases the drag force on you.

**ConnectED** Your one-stop online resource

connectED.mcgraw-hill.com

Video

WebQuest

Audio

Assessment

Review

Concepts in Motion

Inquiry

Multilingual eGlossary

### Reading Guide

**Key Concepts**
**ESSENTIAL QUESTIONS**

- How do force and area affect pressure?

- How does pressure change with depth in the atmosphere and under water?

- What factors affect the density of a fluid?

**Vocabulary**

**fluid** p. 123

**pressure** p. 124

**atmospheric pressure** p. 126

g  **Multilingual eGlossary**

# Pressure and Density of Fluids

## Inquiry  Why Not Straight?

Have you ever noticed that dams are thicker at the bottom than they are at the top? Why do you think people build dams this way?

## What changes? What doesn't?

What happens to the volume of a liquid as you move the liquid from one container to another?

1 Read and complete a lab safety form.

2 As you do each of the following steps, use the markings on the containers to measure the liquid's volume. Record the measurements in your Science Journal.

3 Measure 100 mL of water in a **graduated cylinder.**

4 Pour the water into a **200-mL beaker.**

5 Pour the water into a **square plastic container.** Then pour it back into the graduated cylinder.

**Think About This**

1. What happened to the shape of the water as you moved it from one container to another? What happened to the volume of the water?

2. 🔑 **Key Concept** How do you think the results would be different if you had used another liquid, such as honey? What would happen if you did this experiment with a gas?

## What is a fluid?

When people tell you to drink plenty of fluids when you play a sport, they usually are referring to water or other liquids. It may surprise you to know that gases, such as the oxygen and the nitrogen in the air, are fluids, too. *A* **fluid** *is any substance that can flow and take the shape of the container that holds it.*

### Liquids

You probably have poured milk from a carton into a glass or a bowl. Milk, like other liquids, flows and takes the shape of its container. The volume of a liquid remains the same in any container. When you pour all the milk from a carton into a glass, the milk in the glass occupies the same volume it occupied in the carton. It just has a different shape.

### Gases

What happens when helium gas fills a balloon? If the balloon is long and thin, then the helium atoms flow into and fill the long, thin shape. If the balloon is spherical, then the helium atoms flow into and fill the spherical shape. Like liquids, helium is a fluid. It takes the shape of its container. However, unlike liquids, helium and other gases do not occupy the same volume in different containers. A gas fills its entire container no matter the size of the container.

✓ **Reading Check** How are gases and liquids different?

**WORD ORIGIN** ···········

**fluid**
from Latin *fluidus,* means "flowing"

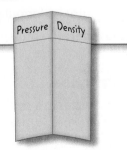
**FOLDABLES®**

Make a vertical two-column chart with labels as shown. Use the chart to organize your notes about pressure and density of fluids.

| Pressure | Density |

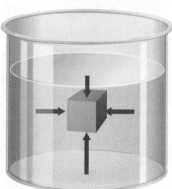

**Figure 1** The pressure applied by a fluid is perpendicular to the surface of any object in contact with the fluid.

# Pressure of Fluids

Have you ever heard someone talk about the air pressure in a car tire or the water pressure on a deep-sea diver? What is pressure? **Pressure** is the amount of force per unit area applied to an object's surface. All fluids, both liquids and gases, apply pressure. The air around you is applying pressure on you right now. Pressure applied on an object by a fluid is related to the weight of the fluid. Like all forces, weight is measured in newtons (N). Pressure can be calculated using the equation below. In the equation, P is pressure, F is the force applied to a surface, and A is the surface area over which the force is applied.

$$\text{pressure} = \text{force} \div \text{area}$$

$$P = \frac{F}{A}$$

The unit measurement for pressure is the pascal (Pa). A dollar bill lying on your hand would apply a pressure of about 1 Pa. A small carton of milk would apply about 1,000 Pa.

## The Direction of Pressure

If the pressure applied by a fluid is related to the weight of the fluid, is the pressure only in the downward direction? No. A fluid applies pressure perpendicular to all sides of an object in contact with the fluid, as shown in **Figure 1.**

## Math Skills ✕⁖ Solve a One-Step Equation

**Solve for Pressure** A surfer on his surfboard applies a force of 645 N on the water. The area of the surfboard is 1.5 m². What is the pressure applied on the water?

**1** **This is what you know:**   force:   $F = 645\text{ N}$
   area:   $A = 1.5\text{ m}^2$

**2** **This is what you need to find:**   pressure:   $P$

**3** **Use this formula:**   $P = \frac{F}{A}$

**4** **Substitute:**
the values for *F* and *A*
into the formula and divide.

$P = \frac{645\text{ N}}{1.5\text{ m}^2}$
$\quad = 430\text{ Pa}$

**Answer:** The pressure is 430 Pa.

• Math Practice
• Personal Tutor

## Practice

The water in a small plastic swimming pool applies a force of 30,000 N over an area of 75 m². What is the pressure on the pool?

### How are force and pressure different?

Sharp pins can pop a balloon. You might think ten pins could pop a balloon more easily than just one pin could. Is this true?

1. Read and complete a lab safety form.

2. Push a **straight pin** through a small piece of **foam board** so the sharp end of the pin extends 1–2 cm.

3. With a partner holding an inflated balloon, gently place the board on the **balloon** so the pin is touching the balloon.

4. Gently apply pressure on the board until the balloon pops.

5. Add nine **more pins** to the foam board. Space them about 0.5 cm apart.

6. While your partner steadies another inflated balloon, hold the board on top of the balloon and gently increase pressure on the board until the balloon pops. Estimate in your Science Journal the difference in pressure you used to pop the balloons.

**Analyze and Conclude**

1. **Analyze** What differences did you notice when trying to pop the balloons?

2. 🔑 **Key Concept** Explain your results by contrasting the force and the pressure per pin on each balloon.

## Pressure and Area

You read on the previous page that the amount of pressure on an object depends on the area over which a force is applied. If you tried to push the box of books shown in **Figure 2** with one finger, your finger would probably bend. It might even hurt. Now suppose you use your entire hand to push the box with the same force. Why is it easier?

When you push a box with one finger, the force you apply spreads over a small area—the surface area of your fingertip. The pressure is great, and your finger hurts. If you push the box with your whole hand, you apply the same force, but the force is spread over a larger area. The pressure is less, and it is easier to push the box.

Pressure decreases when the surface area over which a force is applied increases. Pressure increases when the surface area over which a force is applied decreases.

🔑 **Key Concept Check** How does pressure change as surface area changes?

**Figure 2** 🔑 When a force is applied over a small area, the pressure is greater than when the same force is applied over a larger area.

area = 0.0001 m$^2$

50 N

**Pressure on box = 500,000 Pa**

area = 0.01 m$^2$

50 N

**Pressure on box = 5,000 Pa**

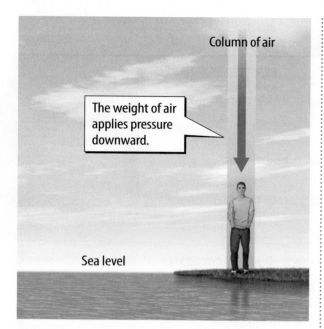

Column of air

The weight of air applies pressure downward.

Sea level

**Figure 3**  The weight of a column of air applies pressure downward. The greater the height above sea level, the lower the pressure.

**REVIEW VOCABULARY**

elevation
  height above sea level

**ACADEMIC VOCABULARY**

sum
  (*noun*) the whole amount

## Pressure and Depth

Have you ever dived under water and felt pressure in your ears? The deeper you dive, the more pressure you feel because pressure applied by a fluid increases with depth.

**Atmospheric Pressure** Think of the air above you as a column that extends high into the atmosphere, as shown in **Figure 3.** The weight of the air in this column applies pressure downward. *The ratio of the weight of all the air above you to your surface area is* **atmospheric pressure.** At sea level, the atmospheric pressure on you and everything around you is about 100,000 Pa, or 100 kPa. Now imagine you are on the top of Mount Everest, nearly 9 km above sea level, as shown in **Figure 4.** Atmospheric pressure is only about 33 kPa. The column of air above you is not as tall as it was at sea level. Therefore, the weight of the air applies less pressure downward. Atmospheric pressure increases as you hike down a mountain, toward lower **elevation.** It decreases as you hike up, to higher elevation.

**Key Concept Check** How does elevation affect atmospheric pressure?

**Underwater Pressure** How does the pressure on the divers in **Figure 4** change as the divers swim deeper? Underwater pressure depends on the **sum** of the weight of the column of air above an object and the weight of the column of water above the object. As a diver dives farther under water, the air column above him or her stays the same. But the water column increases in height, and it weighs more. Underwater pressure increases with depth. At the deepest part of the ocean, the Mariana Trench, the pressure is 108,600 kPa—more than 1,000 times greater than pressure at the ocean's surface. Increased pressure with depth explains why the wall of a dam is thicker at the bottom than at the top. The water behind the dam applies more pressure to the bottom of the wall.

**Figure 4** On land, the atmospheric pressure depends on your elevation. Under water, the water pressure depends on your depth below the water's surface.

✔️ **Visual Check** What would be the pressure on an airplane flying at an altitude of 5,500 m?

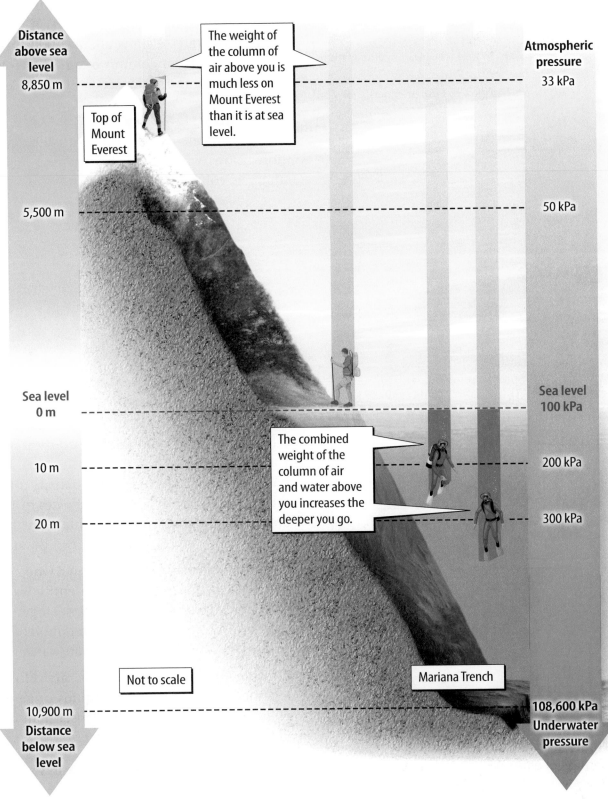

Distance above sea level

8,850 m — Top of Mount Everest

The weight of the column of air above you is much less on Mount Everest than it is at sea level.

5,500 m

Sea level 0 m

10 m

20 m

The combined weight of the column of air and water above you increases the deeper you go.

Not to scale

Mariana Trench

10,900 m

Distance below sea level

Atmospheric pressure

33 kPa

50 kPa

Sea level 100 kPa

200 kPa

300 kPa

108,600 kPa Underwater pressure

SCIENCE USE V. COMMON USE

dense

**Science Use** describes the measure of the ratio of mass to volume

**Common Use** slow to learn or understand

Review    Personal Tutor

**Air**
density = 0.0013 g/cm³

**Ethanol**
density = 0.79 g/cm³

**Vegetable oil**
density = 0.91 g/cm³

**Water**
density = 1.0 g/cm³

**Honey**
density = 1.36 g/cm³

**Figure 5** Materials have different densities because of differences in the masses of their molecules and in the distances between them.

# Density of Fluids

If you walked 10 m down a hill, you would not feel a change in pressure. But if you dived 10 m under water, you would feel a big pressure change. Why is there more pressure under water? The reason is density. If the volume of two fluids is the same, the fluid that weighs more is **denser**. A 10-m column of water weighs about 1,000 times more than a 10-m column of air. Therefore, the water column is denser.

## Calculating Density

You can calculate density with the equation below, where $D$ is density, $m$ is mass, and $V$ is volume.

$$\text{Density} = \text{mass} \div \text{volume}$$

$$D = \frac{m}{V}$$

Density is often measured in grams per cubic centimeter (g/cm³). The density of water is 1.0 g/cm³. The density of air at 100 kPa is about 0.001 g/cm³.

## Densities of Different Materials

Have you ever noticed that vinegar and oil form layers in a bottle of salad dressing? Each layer, such as those shown in **Figure 5,** has a different density. The density of a material is determined by the masses of the atoms or the molecules that make up the material and the distances between them.

A penny is made mostly of zinc. One zinc atom has greater mass than an entire water molecule. Also, zinc atoms are closer together than the atoms are in a water molecule. Therefore, a penny is denser than water.

Molecules in most solids are closer together than molecules in liquids or gases. Therefore, solids are usually denser than liquids or gases.

**Key Concept Check** What factors determine the density of fluids?

# Lesson 1 Review

## Visual Summary

Pressure is high when a force is applied over a small area.

Atmospheric pressure decreases with elevation.

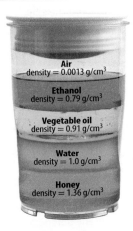

Air
density = 0.0013 g/cm³

Ethanol
density = 0.79 g/cm³

Vegetable oil
density = 0.91 g/cm³

Water
density = 1.0 g/cm³

Honey
density = 1.36 g/cm³

Fluids form layers depending on their densities.

**FOLDABLES**

Use your lesson Foldable to review the lesson. Save your Foldable for the project at the end of the chapter.

## What do you think NOW?

You first read the statements below at the beginning of the chapter.

**1.** Air is a fluid.

**2.** Pressure is a force acting on a fluid.

Did you change your mind about whether you agree or disagree with the statements? Rewrite any false statements to make them true.

## Use Vocabulary

1. **Compare** pressure and atmospheric pressure.

2. Gases and liquids are both _____.

## Understand Key Concepts

3. Ice floats in water. Which must be true?
   A. Ice molecules have more mass than water molecules.
   B. Ice molecules have less mass than water molecules.
   C. A volume of ice has less mass than the same volume of water.
   D. One gram of ice has less volume than 1 g of water.

4. **Compare** One person walks 5 m down a hill. Another dives 5 m in a pool. Which feels a greater pressure increase? Explain.

5. **Draw** a beaker filled with two liquids. One liquid is blue with a density of 0.93 g/cm². The other liquid is red with a density of 1.03 g/cm².

## Interpret Graphics

6. **Organize Information** Copy and fill in the following graphic organizer to list two factors that affect density.

   Density ├─[    ]
           └─[    ]

## Critical Thinking

7. **Propose** As you walk across a field covered with deep snow, you keep sinking. What could you do to prevent this? Explain your reasoning.

## Math Skills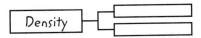

Review
Math Practice

8. What pressure does 50,000 N of water apply to a 0.25-m² surface area of coral?

9. How much pressure is on the bottom of a pot that holds 20 N of soup? The surface area of the pot is 0.05 m²?

# Submersibles

**Deep-sea craft carry scientists to the depths of the ocean.**

Exploring the deep ocean is exciting work. However, the pressure that ocean water applies increases rapidly as depth increases. This pressure limits how far humans can dive without special equipment. One way that humans can explore the deep ocean is to use a submersible.

A submersible is a special watercraft designed to function in deep water. Some submersibles are remote-controlled and do not carry people. Others, such as the *Johnson-Sea-Link* submersible shown below, can carry people. They are designed to protect passengers from pressure in the deep ocean.

A submersible is carried to the dive location on a ship. A lift is used to lower the submersible into the water. At the conclusion of the dive, the submersible is reattached to this lift and hoisted back onto the ship.

HARBOR BRANCH

A steel sphere with an acrylic dome protects the scientists inside from the extreme deep-ocean pressure.

Lights attached to the *JSL* submersible enable scientists to see in the darkness of the deep ocean.

Scientists manipulate robotic arms in order to lift and grab various objects.

This submersible, one of two *Johnson-Sea-Link* submersibles, can transport humans safely to a depth of nearly 1,000 m under water. It carries the oxygen that the pilot and the passengers will need throughout the dive time. The *JSL* submersibles were key to exploring the wreckage of the Civil War ironclad USS *Monitor* off the coast of North Carolina as well as the space shuttle *Challenger* wreckage off the coast of Florida.

## It's Your Turn

**REPORT** Recall that atmospheric pressure decreases with increased elevation. How do airplanes protect people from the decreased air pressure when flying at high elevation? Find out and report what you learn.

### Reading Guide

**Key Concepts**
ESSENTIAL QUESTIONS

- How are pressure and the buoyant force related?

- How does Archimedes' principle describe the buoyant force?

- What makes an object sink or float in a fluid?

**Vocabulary**
**buoyant force** p. 132
**Archimedes' principle** p. 134

g **Multilingual eGlossary**

# The Buoyant Force

**Inquiry  Why doesn't it sink?**

This boat is overloaded. Yet it still floats. What forces are on a boat sitting in water? What forces affect whether it floats or sinks?

## How can objects denser than water float on water?

Many ships are made of aluminum. Aluminum is denser than water. Why does it float?

1. Read and complete a lab safety form.

2. Use **scissors** to cut three 10- × 10-cm squares of **aluminum foil.**

3. Form a boat shape from one square of foil. Squeeze another square into a tight ball. Fold the third square several times into a 2- × 2-cm square. Flatten it completely.

4. Predict whether each object will sink or float. Then, gently place each in a **tub of water.** Record your observations in your Science Journal.

### Think About This

🔑 **Key Concept** What do you think caused each object to float or sink?

## What is a buoyant force?

Have you ever floated on a raft, like the person in **Figure 6?** What forces are acting on this person and his raft? The gravitational force pulls them down. Recall that the gravitational force on an object is the object's weight. The weight of the person and his raft includes the weight of the column of air above them. Why don't they sink? The total weight of the raft, the person, and the air is balanced by another, upward force. *A* **buoyant** (BOY unt) **force** *is an upward force applied by a fluid on an object in the fluid.* The buoyant force on the raft is equal to the total weight of the raft, so the raft floats.

### Buoyant Forces in Liquids

You might have noticed that you can lift someone in water who you could not lift when you are both out of water. This is because of a buoyant force. A buoyant force acts on any object in a liquid. This includes floating objects, such as rafts, ships, and bath toys. It also includes submerged objects, such as rocks at the bottom of a lake.

### Buoyant Forces in Air

Objects in a gas also experience a buoyant force. For example, the buoyant force from air keeps a helium balloon up even though a gravitational force pulls downward. An object does not need to be floating in air for a buoyant force to act on it. In fact, a buoyant force from air is acting on you right now. However, the buoyant force acting on you is less than the gravitational force acting on you. Therefore, you do not float.

**WORD ORIGIN** ············

**buoyant**
from Spanish *boyar,* means "to float"

**Figure 6** A buoyant force balances the weight of this person and his raft and keeps them afloat.

Buoyant force

Weight

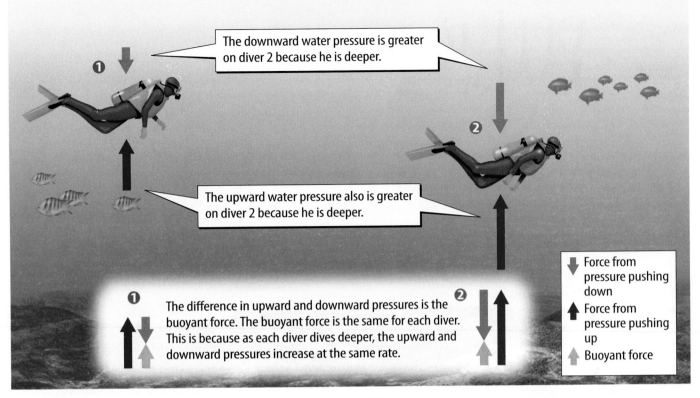

The downward water pressure is greater on diver 2 because he is deeper.

The upward water pressure also is greater on diver 2 because he is deeper.

❶ The difference in upward and downward pressures is the buoyant force. The buoyant force is the same for each diver. This is because as each diver dives deeper, the upward and downward pressures increase at the same rate. ❷

Force from pressure pushing down

Force from pressure pushing up

Buoyant force

**Figure 7**  The buoyant force on a diver is the difference between the force from pressure above and below the diver.

## Buoyant Force and Pressure

How does a fluid, such as water, exert an upward force on any object in the fluid? You read in Lesson 1 that a fluid applies pressure perpendicular to all sides of an object within it. Therefore, the forces from water pressure on the divers in **Figure 7** are in the horizontal and vertical directions. The horizontal forces on the sides of each diver are equal. But, the vertical forces on the top and the bottom of each diver are not equal.

Recall that pressure increases with depth. The pressure at the bottom of each diver is greater than the pressure at the top of each diver. This is illustrated in **Figure 7** by the differences in length of the purple arrows. The dark purple arrow represents the pressure pushing up on each diver. The light purple arrow represents the pressure pushing down on each diver. Notice that the arrows are longer on the diver on the right. That is because this diver is deeper than the other diver. The pressure on both the top and the bottom of this diver is greater.

The difference between the upward force from pressure and the downward force from pressure on each diver is the buoyant force on each diver. The buoyant force is represented by the orange arrows in **Figure 7.** The buoyant force on an object is always in an upward direction because the pressure is always greater at an object's bottom.

**Key Concept Check** How is pressure related to the buoyant force?

## Buoyant Force and Depth

Does the buoyant force on a diver change as she dives deeper? Except with extreme changes of depth, the buoyant force does not significantly change. This is because the pressure of the water above the diver and the pressure below the diver increase by about the same amount as the diver goes deeper. The depth of an object submerged in a fluid has little effect on the buoyant force.

Balloon     Tennis ball     Billiard ball

**Figure 8** 🔑⇀ The buoyant force is greater on the balloon than on the tennis ball or the billiard ball because the balloon displaces more water.

## Archimedes' Principle

There is another way to think about the buoyant force. What happens to each object in **Figure 8** as it is pushed underwater? As each object is submerged in a large beaker, some water spills out, or is displaced, into a smaller beaker. Notice that the balloon displaces more water than the tennis ball or the billiard ball does. The volume of water displaced by the balloon is equal to the volume of the balloon. Similarly, the volume of water displaced by the tennis and the billiard ball is equal to the volume of each ball.

Which small beaker do you think weighs the most? The beaker on the left does because it has the most water in it. According to **Archimedes'** (ar kuh MEE deez) **principle,** *the weight of the fluid that an object displaces is equal to the buoyant force acting on the object.* Because the balloon displaces more water than the billiard ball or the tennis ball does, a greater buoyant force acts on it.

🔑⇀ **Key Concept Check** What is the buoyant force on you if you displace 400 N of water as you dive under water?

Recall the image of the divers on the previous page. The buoyant force on each diver does not change as the divers go deeper because the volume of each diver does not change. Each displaces the same amount of water at any depth.

**Weight is greater than buoyant force.**

**Buoyant force and weight are balanced.**

⬆ Buoyant force
⬇ Weight

## Sinking and Floating

Now look at the tennis ball and the billiard ball in **Figure 9.** Both have the same volume. But, if you set each on the surface of water and let go, one floats and one sinks.

### Buoyant Force and Weight

If you set a tennis ball on the surface of water, it displaces water. With only a little bit of the tennis ball submerged, the tennis ball stops displacing water and floats. Why is this? When the weight of the water displaced by an object equals the weight of the object, the object floats. So, the weight of the water displaced by the tennis ball, or the buoyant force, equals the weight of the tennis ball. The same is true for the leaf in **Figure 9.**

However, when you set a billiard ball on the surface of water, it sinks. Even after the billiard ball is completely submerged and has displaced the maximum volume of water that it can, it still weighs more than the water it displaces. When an object's weight is greater than the buoyant force, the object sinks.

The buoyant force on the billiard ball is greater than the buoyant force on the tennis ball because the billiard ball is completely submerged. The water displaced by the sunken billiard ball weighs more than the water displaced by the floating tennis ball.

### Buoyant Force and Density

Because the billiard ball sinks, you know it weighs more than the water it displaces. Therefore, the billiard ball has more mass per volume, or a greater density, than water. If an object is more dense than the fluid in which it is placed, then the buoyant force on that object will be less than the object's weight, and the object will sink.

**Figure 9** If the weight of an object is greater than the buoyant force acting on it, the object sinks.

✅ **Visual Check** How do the buoyant forces on the billiard ball, the tennis ball, and the leaf compare?

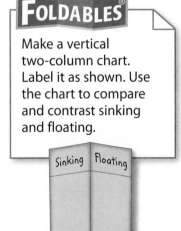

**FOLDABLES**

Make a vertical two-column chart. Label it as shown. Use the chart to compare and contrast sinking and floating.

Sinking | Floating

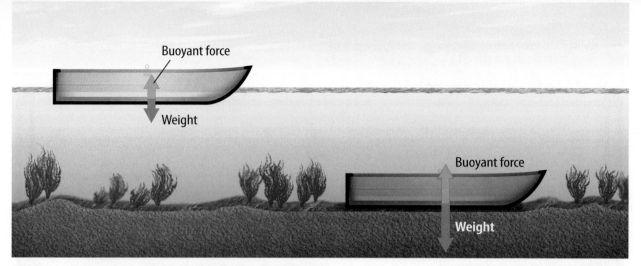

 **Figure 10** 🔑 The boat on the left floats because it is filled with air instead of water.

**Figure 11** 🔑 As a balloon loses helium, its density increases and its buoyant force decreases.

🔘 **Visual Check** How would the forces be different on the balloon if it were neither rising nor falling but was floating in the air? ▼

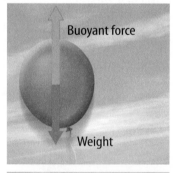

## Metal Boats

Aluminum has more than twice the density of water. So, how does the aluminum boat in the left part of **Figure 10** stay afloat?

The boats in **Figure 10** are the same size and made from the same material. The boat on the left is filled with air. The weight of the water displaced equals the weight of the boat plus the air inside. Therefore, the boat floats.

The boat on the right is filled with water. In order for this boat to float, the buoyant force would need to equal the weight of the boat plus the water inside. However, the total weight of the boat and the water inside is greater than the water displaced by the boat, or the buoyant force. So, the boat sinks.

You can also think of this is in terms of density. The density of the boat and air on the left is less than the density of water, so the boat floats. However, the density of the boat and water on the right is greater than the density of water. It sinks. The buoyant force is still greater on the boat on the right because it displaces more water than the boat on the left.

🔑 **Key Concept Check** If an object weighing 14 N experiences a 12-N buoyant force, will it sink or float?

## Balloons

Why does a helium balloon rise? Helium is less dense than either oxygen or nitrogen in the air. Therefore, the buoyant force acting on a balloon filled with helium is greater than the weight of the balloon, as illustrated in the top of **Figure 11.** The buoyant force pushes the balloon upward. After a day or two, a helium balloon begins to shrink. Helium atoms are so small that they pass between particles that make up the balloon. As the volume of the balloon decreases, the density of the balloon increases. Eventually, the balloon's density is greater than the air's density. The buoyant force on the balloon is less than the balloon's weight, and the balloon falls to the ground.

# Lesson 2 Review

## Visual Summary

A buoyant force results from the difference in pressure between the top and the bottom of an object.

Objects that have the same volume in a fluid experience the same buoyant force.

Buoyant force

Weight

When the density of a balloon becomes greater than the density of air, the balloon sinks.

**FOLDABLES**

Use your lesson Foldable to review the lesson. Save your Foldable for the project at the end of the chapter.

## What do you think NOW?

You first read the statements below at the beginning of the chapter.

**3.** You can lift a rock easily under water because there is a buoyant force on the rock.

**4.** The buoyant force on an object depends on the object's weight.

Did you change your mind about whether you agree or disagree with the statements? Rewrite any false statements to make them true.

## Use Vocabulary

**1** **State** Archimedes' principle in your own words.

**2** **Relate** buoyant force to pressure.

## Understand Key Concepts

**3** **Construct** a diagram of the forces on a jellyfish suspended in water. Include forces from pressure on the top, the sides, and the bottom of the jellyfish. Also include the buoyant force and the force from the weight of the jellyfish.

**4** The density of water is 1.0 g/cm$^3$. Which of these floats in water?
  A. a ball with a density of 1.7 g/cm$^3$
  B. a ruler with a density of 1.1 g/cm$^3$
  C. a cup that weighs 2 N and displaces 3 N of water when submerged
  D. a toy that weighs 4 N and displaces 3 N of water when submerged

## Interpret Graphics

**5** **Identify** If the person and the raft shown below together weigh 550 N, then how large is the buoyant force on them?

**6** **Organize Information** Copy and complete the graphic organizer stating the relative sizes of forces acting on a sinking object.

Object sinks

## Critical Thinking

**7** **Use** Archimedes' principle to explain why it is easier to lift a rock in water than it is to lift the same rock in air.

# Do heavy objects always sink and light objects always float?

You have seen many objects that sink in water and many that float. Is it possible to predict whether an object will sink or float if you know its mass? In this lab, you will measure the mass of various objects and then predict whether the objects sink or float.

## Materials

triple-beam balance

craft stick

paper towels

tub

**Also needed:**
Objects: sink or float?

### Safety

### Learn It

When you **predict** the results of a scientific investigation, you tell what you think will happen. You should base your prediction on what you already know and on things you observe.

### Try It

1. Read and complete a lab safety form.

2. Copy the data table below into your Science Journal. Add more rows as needed.

3. Measure the mass of a craft stick. Record it in your data table.

4. Predict whether the craft stick will sink or float. Record your prediction.

5. Place the craft stick in the tub of water. Does it sink, or does it float? Record your results.

### Apply It

6. Measure and record the mass of each of the other objects.

7. Predict whether each object will sink or float. Record your predictions.

8. Place each object in the water, and observe whether it sinks or floats. Record your results.

9. 🔑 **Key Concept** Do heavy objects always sink and light objects always float? Explain your reasoning.

| Object | Mass (g) | Predict Sink or Float? | Observe Sink or Float? |
|--------|----------|------------------------|------------------------|
|        |          |                        |                        |
|        |          |                        |                        |
|        |          |                        |                        |

# Other Effects of Fluid Forces

## Reading Guide

### Key Concepts
### ESSENTIAL QUESTIONS

- How are forces transferred through a fluid?

- How does Bernoulli's principle describe the relationship between pressure and speed?

- What affects drag forces?

### Vocabulary

**Pascal's principle** p. 140

**Bernoulli's principle** p. 142

**drag force** p. 144

**g** Multilingual eGlossary

## Inquiry Why the Parachute?

Have you ever flown in an airplane and noticed that the wing flaps move up as the plane prepares to land? The flaps increase the drag force on the plane, slowing it down. Similarly, the parachute attached to this race car increases the drag force on the car and slows it down.

## How is force transferred through a fluid?

Fluids apply forces. How are forces transferred through fluids?

1. Read and complete a lab safety form.

2. With a partner, use a **nail** to carefully poke three small, identical holes about 1 cm apart along one side of a **drinking straw.** Use **tape** to connect one end of this straw to one end of a **bendable straw.**

3. Tape a **funnel** to the open end of the bendable straw.

4. Keeping the straws and funnel connected, place them in a **tub.** The straw with the holes should be horizontal, with the holes facing upward. Have your partner hold a finger over the open end of the horizontal straw.

5. Gently pour water into the funnel. Record your observations in your Science Journal.

### Think About This

1. How did the streams of water pouring out of the holes differ?

2. 🔑 **Key Concept** How do you think force from pressure is transferred through a fluid?

## Fluid Forces—Benefits and Challenges

You use forces in fluids to accomplish everyday tasks. You produce a force when you water a garden with a hose or squeeze a plastic ketchup bottle. You make use of a buoyant force when you float on a raft or lift a rock under water.

Forces in fluids also can make tasks difficult or even dangerous. When a car moves at high speeds, air pushes against the car. This increases the amount of fuel the car uses. Fluid forces from floods, tornadoes, and hurricanes can cause damage that costs lives and billions of dollars. How are forces in fluids both useful and dangerous?

## Pascal's Principle

Blaise Pascal (pas KAL), a 17th century French physicist, studied the pressures of fluids in closed containers. A fluid cannot flow into or out of a closed container. **Pascal's principle** *states that when pressure is applied to a fluid in a closed container, the pressure increases by the same amount everywhere in the container.* If you push on a closed ketchup bottle, the pressure on the ketchup increases not only under your fingers, it increases equally throughout the bottle.

🔑 **Key Concept Check** How does pressure change when force is applied to a fluid in a closed container?

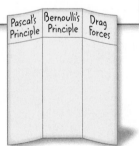

**FOLDABLES®**

Make a horizontal three-column chart book. Label it as shown. Use it to organize your notes on Pascal's principle, Bernoulli's principle, and drag forces.

## Pushing on a Fluid

Pascal's principle applies to fluid power systems. A fluid power system like the one in **Figure 12** uses a pressurized fluid to transfer motion. The figure shows the piston on the left moving down as the piston on the right moves up. In this example, the surface area of the input piston is half that of the output piston. Any input force applied to the input piston results in an output force that is doubled by the output piston.

Why is the output force greater than the input force? Pascal's principle can be written as $F = P \times a$. If pressure ($P$) is always equal throughout the system, increasing the area of the output piston ($a$) results in a larger force ($F$). Does the output piston do more work than the input piston?

No. Recall that *work = force × distance*. In this example, the output force is twice that of the input force. And, if you were to build the fluid power system shown in **Figure 12**, you would find that the input piston is pushed down twice the distance the output piston moves up. If you use the work equation to solve for the input work and then for the output work of the system, you would find that the two are equal.

## Hydraulic Lifts

Automobile mechanics use Pascal's principle to lift a car with a hydraulic lift, as shown in **Figure 13**. A hydraulic lift is an example of a fluid power system that uses a liquid for the fluid. In an oil-filled hydraulic lift, a narrow tube connects to a wider tube under a car. Pushing down on a piston in the narrow tube generates an upward force on a larger piston great enough to lift a car.

Like the fluid power system in **Figure 12**, a hydraulic lift does not reduce the amount of work needed to lift a car. To lift a car, the piston in the narrow tube travels a longer distance than the piston in the wider tube.

✔ **Reading Check** On what principle do hydraulic lifts rely?

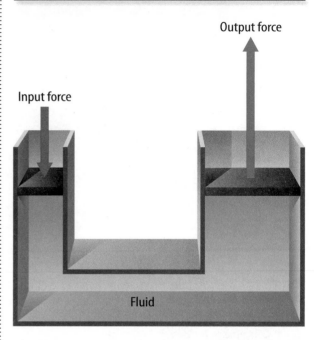

**Pascal's principle** 🔑

Output force

Input force

Fluid

▲ **Figure 12** The left side of the piston has a surface area that is half that of the right side. A force pushing down on the left side generates a force twice as large on the right side.

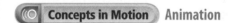

**Visual Check** If you push down with a force of 100 N on the left side of the piston, what is the force on the right side?

**Concepts in Motion** Animation

**Figure 13** People rely on Pascal's principle when they use hydraulic lifts. A hydraulic lift increases the force lifting up on a car. ▼

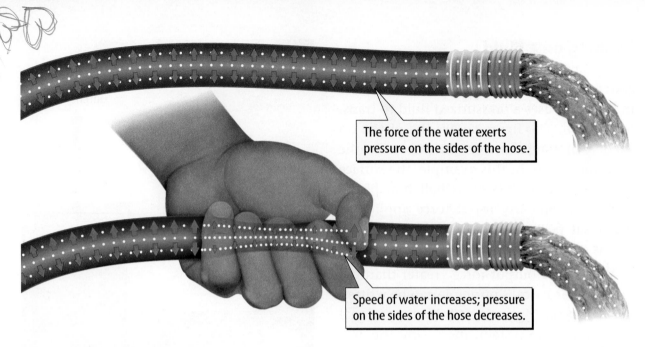

The force of the water exerts pressure on the sides of the hose.

Speed of water increases; pressure on the sides of the hose decreases.

**Figure 14** According to Bernoulli's principle, when water moves faster in the pinched part of a hose, it applies less pressure on the sides of the hose.

**Concepts in Motion** Animation

### How does air speed affect air pressure?

What happens when moving air changes the pressure on an object?

1. Read and complete a lab safety form.

2. Use **tape** to attach a 25-cm **string** to each of two empty **aluminum cans.** Hold the cans 1–2 cm apart in front of your face.

3. Gently blow between the cans. Record your observations in your Science Journal.

**Analyze and Conclude**

1. **Describe** how the cans moved.

2. **Key Concept** Explain how this activity demonstrates Bernoulli's principle.

## Bernoulli's Principle

Have you ever sprayed a friend with water from a hose? Garden hoses illustrate a principle that describes the relationship between speed and pressure in fluids. **Bernoulli's** (ber NEW leez) **principle** *states that the pressure of a fluid decreases when the speed of that fluid increases.*

Consider the hose shown at the top of **Figure 14.** As the water flows through the hose, it applies a constant pressure on the sides of the hose, and it comes out of the hose at a constant speed.

Suppose you pinch the hose slightly, as shown in the bottom of **Figure 14.** The water speeds up in the pinched part. After it has traveled through the pinched part, it returns to its regular speed. According to Bernoulli's principle, when speed increases, pressure decreases. This means the pressure of the water on the sides of the hose in the pinched part is less than it is elsewhere in the hose.

 **Key Concept Check** What is the relationship between speed and pressure in a fluid?

## Damage from High Winds

Have you ever seen a photograph of a house that has had its roof blown off during a windstorm? When strong winds blow over a house, the air outside the house has a high speed. But the air inside the house has almost no speed. According to Bernoulli's principle, increased speed means lowered pressure. Therefore, the pressure outside the house is lower than the pressure inside the house. This is illustrated in the left side of **Figure 15.** The force of the air pressure pushing down on the roof of this house is less than the force of the air pressure pushing up. When the upward force inside the house becomes greater than the combined downward force outside the house— including the force of gravity—the roof begins to rise.

 **Key Concept Check** How does Bernoulli's principle explain how wind can take the roof off a house?

Now think about what happens immediately after the roof lifts off the house, as shown in the right side of **Figure 15.** The wind continues to blow. Its speed is now the same below and above the roof. The air pressure is also the same below and above the roof. The force from the upward air pressure on the roof is balanced by the force from the downward air pressure on the roof. But, the force of gravity adds to the force from the downward air pressure. When the combined downward force is greater than the upward force, the roof crashes back down.

**Figure 15** When the upward pressure on a roof is greater than the downward pressure, the roof moves up.

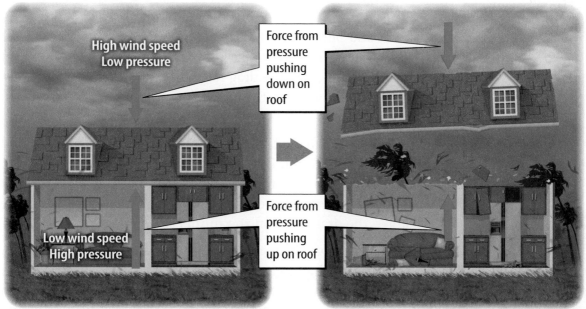

If wind is moving very quickly outside a house, such as during a tornado, the pressure outside will be less than the pressure inside. If the pressure inside the house is greater than the pressure outside, the roof can lift off.

If a roof lifts off a house, the pressure outside the house pushing down on the roof and the pressure inside the house pushing up on the roof are equal.

Lower wind speed
Higher pressure

Higher wind speed
Lower pressure

▲ **Figure 16** The higher pressure on the left side of the soccer ball causes the ball to curve right.

✔ **Visual Check** Which way would the ball curve if it were spinning in the opposite direction?

**WORD ORIGIN** · · · · · · · · · ·

**drag force**
from Old Norse *draga,* means "to draw"; and Latin *fortis,* means "force"

**Figure 17** 🔑 The parachute increases the drag force on this runner and makes him work harder. ▼

## Soccer Kicks

You might have seen a soccer player kick a soccer ball so that it curves around an opposing player. A soccer player who curves the ball this way is making use of Bernoulli's principle.

The soccer player puts a spin on the ball as he or she kicks it, as shown in **Figure 16.** This makes the speed of the air on one side of the ball greater than the speed of the air on the other side. According to Bernoulli's principle, the side with the lower speed has greater air pressure acting on it. Because air moves from areas of high to low pressure, the spinning ball curves toward the side with lower pressure.

## Drag Forces

Have you ever seen a runner using a parachute, like in **Figure 17?** Runners use parachutes to increase the drag force on them and build strength. The **drag force** is a force that opposes the motion of an object through a fluid. As the speed of an object in a fluid increases, the drag force on that object also increases. The faster the runner, the greater the drag force on him or her, and the harder he or she needs to work to resist it.

The drag force on an object also depends on the size and shape of that object. If two objects move in the same direction, the object with the greater surface area toward the direction of motion has a greater drag force on it. If the parachute were larger, the drag force on the runner would be even greater.

The drag force increases when the density of a fluid increases. You probably have felt the pull of water against your legs when you wade in shallow water. When you walk through air, which has a lower density, you barely notice the drag force. Whether you notice them or not, the drag force and other forces in fluids are all around you. They help you work and play.

🔑 **Key Concept Check** What affects the drag force on an object?

## Visual Summary

People rely on Pascal's principle when they use hydraulic lifts.

The imbalance of pressures in fluids can cause a roof to lift off a house in a severe windstorm.

A soccer player who kicks a curved ball makes use of Bernoulli's principle.

**FOLDABLES**

Use your lesson Foldable to review the lesson. Save your Foldable for the project at the end of the chapter.

## What do you think NOW?

You first read the statements below at the beginning of the chapter.

**5.** If you squeeze an unopened plastic ketchup bottle, the pressure on the ketchup changes everywhere in the bottle.

**6.** Running with an open parachute decreases the drag force on you.

Did you change your mind about whether you agree or disagree with the statements? Rewrite any false statements to make them true.

## Use Vocabulary

1 **State** Bernoulli's principle in your own words.

2 The transfer of forces in fluids in closed containers is explained by _____.

3 **Use the term** *drag force* in a sentence.

## Understand Key Concepts 🔑

4 **Identify** Water flows through one pipe section at 4 m/s and then through a second section at 8 m/s. In which section is the pressure greater? Explain why.

5 **Determine** A closed container of liquid soap is 10 cm tall. You squeeze on the container's top with a pressure of 1,000 Pa. What is the pressure at the bottom?

6 **Explain** how drag force differs on two cars traveling at different speeds. Why does the car traveling faster use more fuel?

## Interpret Graphics

7 **Analyze** In the diagram below, if you push the left side down 3 m, how far will the right side move up?

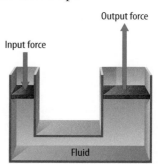

Output force

Input force

Fluid

8 **Organize Information** Copy and fill in the graphic organizer below to list four things that affect drag forces.

Drag forces

## Critical Thinking

9 **Design** a hydraulic lift that could lift 100 N 1 m with a 25-N force.

## Materials

craft supplies

creative building materials

office supplies

rocks

## Safety

# Design a Cargo Ship

Modern cargo ships can carry thousands of freight containers. But cargo ships will sink if they are loaded with too much cargo. In this lab, you will build a model cargo ship. You then will test the ship's buoyancy to determine how much weight it can support.

## Question

How is buoyant force related to weight in a model cargo ship?

## Procedure

1. Read and complete a lab safety form.
2. Look at the available supplies. Discuss with your partner how you can build a cargo ship using the supplies.
3. Consider what you have learned about the buoyant force and about Archimedes' principle. What features should your boat have to enable it to carry heavy cargo?
4. Work with your partner to build your cargo ship.
5. Measure the weight of your ship. Record it in your Science Journal in newtons. Use this formula to convert grams to newtons:

    N = mass (in kg) × 9.8
6. Place the ship in a large tub of water. How high on the water does the ship float? How much of the ship is below water?

7. Add rocks to your ship to model freight containers. Before you add a rock, measure it and record its weight in your Science Journal. Note how the position of the ship on the water changes with each rock.

8. Continue adding rocks until your ship sinks. Then, calculate the total weight of the ship and the rocks it was able to support before it sank. Show your calculations in your Science Journal.

## Analyze and Conclude

9. **Measure** What is the weight of your ship?

10. **Measure** What was the total weight of the ship and the rocks before it sank?

11. **Assess** How did the buoyant force on your ship change each time you added a rock?

12. **The Big Idea** Describe how the buoyant force kept your ship afloat.

13. **Relate** How did the buoyant force on your ship relate to its volume?

14. **Interpret** How can you explain your results in terms of density?

## Communicate Your Results

With your partner, create a brochure advertising your ship. Include how much weight it can carry. Compare brochures with those of your classmates. How did the ability of your ship to support weight compare with the ships of other groups? What was different about the ships that could support the most weight?

**Inquiry Extension**

Place the small tub of water in a larger tub. Fill the small tub completely with water. Place two rocks on your ship, and set the ship on the water. Measure the weight of the water it displaces. Fill the small tub again. Add to the ship all the rocks your ship can support. Again measure the weight of the water it displaces. How did the buoyant force change? In your Science Journal, draw your ship with no rocks, with two rocks, and with all the rocks it could support. Label all forces on the ship in each drawing.

### Lab Tips

☑ As you design your ship, think about how you can maximize the amount of water displaced by your ship while keeping the weight of the ship to a minimum.

☑ When calculating the total weight supported by your ship, be sure not to include the weight of the final rock that sank the ship.

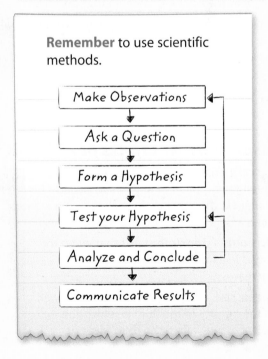

**Remember** to use scientific methods.

Make Observations
↓
Ask a Question
↓
Form a Hypothesis
↓
Test your Hypothesis
↓
Analyze and Conclude
↓
Communicate Results

People use forces in fluids to float objects on water and in air, to lift objects, and to affect the motions of objects.

## Key Concepts Summary

| | |

### Lesson 1: Pressure and Density of Fluids

- **Pressure** is the ratio of force to area.
- **Atmospheric pressure** decreases with elevation. Pressure under water increases with depth.
- The density of a **fluid** depends on the mass of the fluid and its volume.

area = 0.0001 m²

### Lesson 2: The Buoyant Force

- The change in pressure between the top and the bottom of an object results in an upward force called the **buoyant force.**
- **Archimedes' principle** states that the weight of the fluid displaced by an object is equal to the buoyant force on that object.
- An object sinks if its weight is greater than the buoyant force on it. An object does not sink if the buoyant force on it is equal to its weight.

Buoyant force

Weight

### Lesson 3: Other Effects of Fluid Forces

- **Pascal's principle** states that when pressure is applied to a fluid in a closed container, the pressure increases by the same amount everywhere in the container.
- **Bernoulli's principle** states that when the speed in a fluid increases, the pressure decreases.
- Speed, size, and shape of an object, as well as the density of the fluid in which the object moves, affect the **drag force** on that object.

## Vocabulary

**fluid** p. 123
**pressure** p. 124
**atmospheric pressure** p. 126

**buoyant force** p. 132
**Archimedes' principle** p. 134

**Pascal's principle** p. 140
**Bernoulli's principle** p. 142
**drag force** p. 144

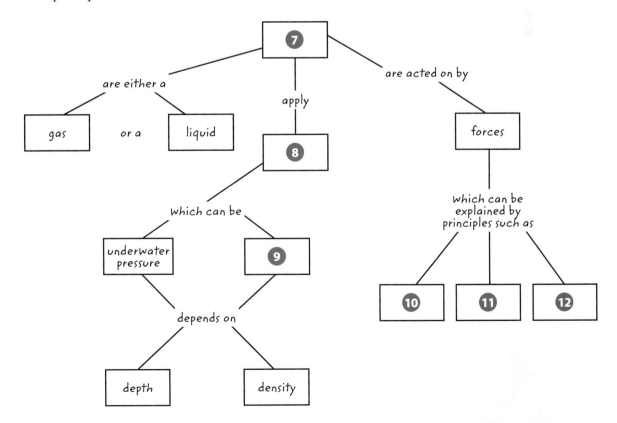
## FOLDABLES® Chapter Project

Assemble your lesson Foldables as shown to make a Chapter Project. Use the project to review what you have learned in this chapter.

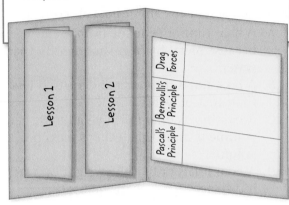

## Use Vocabulary

1 Air and water are both _____.

2 A ship does not sink because a(n) _____ acts on it.

3 A column of air exerts _____ on you.

4 According to _____, two objects of equal volume in a fluid experience the same buoyant force.

5 A parachute with a 5-m² surface area experiences a much larger _____ than a parachute with a 3-m² surface area.

6 Fluid power systems work according to _____.

## Link Vocabulary and Key Concepts

Concepts in Motion    Interactive Concept Map

Copy this concept map, and then use vocabulary terms from the previous page to complete the concept map.

## Understand Key Concepts

**1** Which is NOT a fluid?
   A. helium
   B. ice
   C. milk
   D. water

**2** If you poured the following fluids into a container, which would float on top?
   A. maple syrup, with a density of 1.33 g/cm³
   B. olive oil, with a density of 0.9 g/m³
   C. seawater, with a density of 1.03 g/cm³
   D. water, with a density of 1.0 g/cm³

**3** What pressure does Adam apply to a ball of dough when he pushes on it with a 25-N force? The area of his hand is 0.01 m².
   A. 0.0004 Pa
   B. 0.5 Pa
   C. 25 Pa
   D. 2,500 Pa

**4** Which of these has the greatest pressure applied to it from the surrounding fluid?
   A. a fish swimming 20 m below the surface
   B. a hawk flying 300 m above sea level
   C. a mountain climber at an altitude of 4,400 m
   D. a person fishing off the coast of California

**5** In the diagram of the beach ball floating on water below, what force does the blue arrow represent?

   A. pressure
   B. weight
   C. buoyant force
   D. drag force

**6** Joseph weighs 290 N and displaces 300 N of water as he swims under water in a pool. What is the buoyant force on Joseph?
   A. 10 N upward
   B. 300 N upward
   C. 290 N downward
   D. 590 N downward

**7** Which statement about boats is correct?
   A. A boat cannot be made from metal because metal has a greater density than water.
   B. A boat floats if its overall density is less than that of water.
   C. A boat floats only if its overall mass per volume is more than water's mass per volume.
   D. A boat floats only if the weight of water it displaces is less than the boat's weight.

**8** In the diagram below, how large a force is applied by the piston on the right?

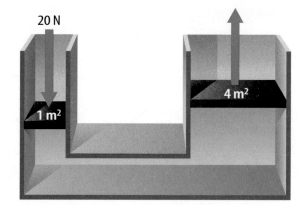

   A. 10 N
   B. 20 N
   C. 40 N
   D. 80 N

**9** A leaf enters a drainpipe, and the pressure in the water increases from 10 Pa to 30 Pa. What happens to the leaf's speed?
   A. It decreases.
   B. It increases.
   C. It becomes exactly 30 m/s.
   D. It does not change.

## Critical Thinking

**10 Design an Experiment** You are given a material and asked to determine whether it is a fluid. What experiment could you do?

**11 Propose** The legs of your chair sink into the sand on a beach. What could you do to prevent this? Explain your reasoning.

**12 Compare** You drop a solid cube into water. The cube is 40 cm on each side. Then you drop another solid object into water that is 160 cm tall, 20 cm deep, and 20 cm wide and made from the same material as the cube. Which object experiences the greater buoyant force? Explain.

**13 Interpret** A sailor drops an anchor over the side of a ship. When the anchor is 10 m below the ocean's surface, the buoyant force on the anchor is 80 N. What is the buoyant force on the anchor when it sinks to 100 m below the surface?

**14 Assess** How does opening a parachute change the drag force on a skydiver?

**15 Explain** Woodchucks live in underground tunnels such as the one shown below. One opening has a dirt mound around it, and air flows across it quickly. The other opening is even with the ground. The air moves across it with less speed. How does this design help ventilate the tunnel?

Air speed

Air speed

*Writing in Science*

**16 Write** a paragraph explaining why the *Titanic* sank. Use what you have learned about forces and fluids.

## REVIEW THE BIG IDEA

**17** The braking systems in most automobiles rely on a hydraulic fluid power system. Use Pascal's principle to explain how the pressure of a foot on a car's brake can stop the car.

**18** The photo below shows a helium balloon in a parade. Imagine you are holding a rope attached to this balloon. What forces do you encounter? Explain at least two ways in which you might encounter or use forces in fluids in your everyday life.

## Math Skills

**Review**

Math Practice

### Solve a One-Step Equation

**19** The buoyant force on an inflatable pool raft with a surface area of 2 m² is 36 N. How large is the upward pressure on the raft?

**20** A ballerina stands on her toes in a pointe shoe. The pressure on her toes is 454,000 Pa. When she stands on flat feet, the pressure is 22,700 Pa. If the ballerina's weight is 454 N, what is the surface area she stands on when she's on her toes, and what is the surface area she stands on when she's flat-footed?

*Record your answers on the answer sheet provided by your teacher or on a sheet of paper.*

## Multiple Choice

**1** The same force is applied over two areas that differ in size. Which is true of the pressure over these areas?

   **A** The pressure is equal to the force multiplied by the area.

   **B** The pressure on both areas is the same.

   **C** The pressure on the larger area is greater.

   **D** The pressure on the smaller area is greater.

*Use the figure to answer question 2.*

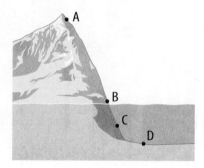

**2** The figure shows a side view of a mountain that extends below the water's surface. At which point would the pressure be 100 kPa?

   **A** A

   **B** B

   **C** C

   **D** D

**3** Which characteristics determine the density of a fluid?

   **A** mass and energy of the particles

   **B** mass of particles and distance between them

   **C** number of particles and distance between them

   **D** shape and energy of the particles

**4** Which changes as a solid object moves downward within a fluid?

   **A** the buoyant force acting on the object

   **B** the mass of the object

   **C** the pressure acting on the object

   **D** the volume of the object

**5** Which explains how the buoyant force on an object changes with the weight of the fluid it displaces?

   **A** Archimedes' principle

   **B** definition of density

   **C** definition of pressure

   **D** Pascal's principle

*Use the figure to answer question 6.*

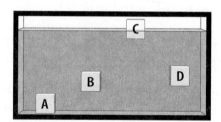

**6** The figure shows an aquarium filled with water. Four cubes made out of different materials have been placed in the aquarium. For which object is the buoyant force acting on it equal to the object's weight?

   **A** A

   **B** B

   **C** C

   **D** D

**7** Why are drag forces greater in water than they are in air?

   **A** Air is denser than water.

   **B** Air is a fluid, but water is not.

   **C** Water is denser than air.

   **D** Water is a fluid, but air is not.

*Use the figure to answer questions 8 and 9.*

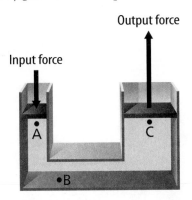

Output force

Input force

8 When an input force is applied above point A, which is true of the change in fluid pressure at points A, B, and C?

  A The fluid pressure increases the most at point A.

  B The fluid pressure increases the most at point B.

  C The fluid pressure increases the most at point C.

  D The fluid pressure increases by the same amount at all three points.

9 Which describes how the input force affects the fluid pressure at different points in the hydraulic lift?

  A Bernoulli's principle

  B definition of buoyant force

  C equation for density

  D Pascal's principle

## Constructed Response

10 Explain how the buoyant force on an underwater diver is exactly the same at two different depths.

11 Use the terms *buoyant force* and *weight* to describe how a helium balloon floats in air.

*Use the figure to answer question 12.*

12 The figure shows two strips of paper held above an air source. When the air jet is turned on, the air between the pieces of paper will move quickly. The air on either side of each piece of paper will hardly move at all. When the air jet is turned on, how will the pressure between the pieces of paper compare to the pressure on the outer sides of each piece? Predict what will happen to the pieces of paper when the air jet is turned on.

13 Mercury is a fluid that has a density of 13.53 g/cm³. Gold is a solid that has a density of 19.32 g/cm³. Aluminum is a solid that has a density of 2.7 g/cm³. Predict what will happen if 5-g samples of gold and aluminum are dropped into a flask of mercury.

| NEED EXTRA HELP? | | | | | | | | | | | | | |
|---|---|---|---|---|---|---|---|---|---|---|---|---|---|
| If You Missed Question... | 1 | 2 | 3 | 4 | 5 | 6 | 7 | 8 | 9 | 10 | 11 | 12 | 13 |
| Go to Lesson... | 1 | 1 | 1 | 2 | 2 | 2 | 3 | 3 | 3 | 2 | 2 | 3 | 1 |

# Student Resources

## For Students and Parents/Guardians

These resources are designed to help you achieve success in science. You will find useful information on laboratory safety, math skills, and science skills. In addition, science reference materials are found in the Reference Handbook. You'll find the information you need to learn and sharpen your skills in these resources.

# Table of Contents

# Scientific Methods

Scientists use an orderly approach called the scientific method to solve problems. This includes organizing and recording data so others can understand them. Scientists use many variations in this method when they solve problems.

## Identify a Question

The first step in a scientific investigation or experiment is to identify a question to be answered or a problem to be solved. For example, you might ask which gasoline is the most efficient.

## Gather and Organize Information

After you have identified your question, begin gathering and organizing information. There are many ways to gather information, such as researching in a library, interviewing those knowledgeable about the subject, and testing and working in the laboratory and field. Fieldwork is investigations and observations done outside of a laboratory.

**Researching Information** Before moving in a new direction, it is important to gather the information that already is known about the subject. Start by asking yourself questions to determine exactly what you need to know. Then you will look for the information in various reference sources, like the student is doing in **Figure 1.** Some sources may include textbooks, encyclopedias, government documents, professional journals, science magazines, and the Internet. Always list the sources of your information.

**Figure 1** The Internet can be a valuable research tool.

**Evaluate Sources of Information** Not all sources of information are reliable. You should evaluate all of your sources of information, and use only those you know to be dependable. For example, if you are researching ways to make homes more energy efficient, a site written by the U.S. Department of Energy would be more reliable than a site written by a company that is trying to sell a new type of weatherproofing material. Also, remember that research always is changing. Consult the most current resources available to you. For example, a 1985 resource about saving energy would not reflect the most recent findings.

Sometimes scientists use data that they did not collect themselves, or conclusions drawn by other researchers. This data must be evaluated carefully. Ask questions about how the data were obtained, if the investigation was carried out properly, and if it has been duplicated exactly with the same results. Would you reach the same conclusion from the data? Only when you have confidence in the data can you believe it is true and feel comfortable using it.

**Interpret Scientific Illustrations** As you research a topic in science, you will see drawings, diagrams, and photographs to help you understand what you read. Some illustrations are included to help you understand an idea that you can't see easily by yourself, like the tiny particles in an atom in **Figure 2.** A drawing helps many people to remember details more easily and provides examples that clarify difficult concepts or give additional information about the topic you are studying. Most illustrations have labels or a caption to identify or to provide more information.

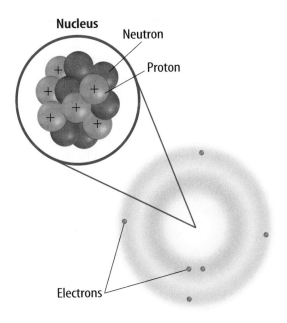

**Figure 2** This drawing shows an atom of carbon with its six protons, six neutrons, and six electrons.

**Concept Maps** One way to organize data is to draw a diagram that shows relationships among ideas (or concepts). A concept map can help make the meanings of ideas and terms more clear, and help you understand and remember what you are studying. Concept maps are useful for breaking large concepts down into smaller parts, making learning easier.

**Network Tree** A type of concept map that not only shows a relationship, but how the concepts are related is a network tree, shown in **Figure 3.** In a network tree, the words are written in the ovals, while the description of the type of relationship is written across the connecting lines.

When constructing a network tree, write down the topic and all major topics on separate pieces of paper or notecards. Then arrange them in order from general to specific. Branch the related concepts from the major concept and describe the relationship on the connecting line. Continue to more specific concepts until finished.

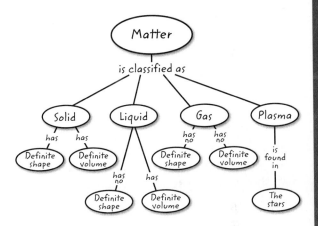

**Figure 3** A network tree shows how concepts or objects are related.

**Events Chain** Another type of concept map is an events chain. Sometimes called a flow chart, it models the order or sequence of items. An events chain can be used to describe a sequence of events, the steps in a procedure, or the stages of a process.

When making an events chain, first find the one event that starts the chain. This event is called the initiating event. Then, find the next event and continue until the outcome is reached, as shown in **Figure 4** on the next page.

SCIENCE SKILL HANDBOOK

MATH SKILL HANDBOOK

FOLDABLES HANDBOOK

REFERENCE HANDBOOK

GLOSSARY/ GLOSARIO

INDEX

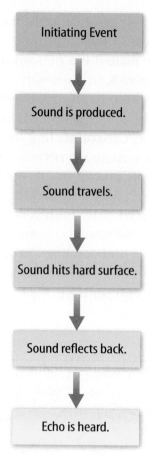

**Figure 4** Events-chain concept maps show the order of steps in a process or event. This concept map shows how a sound makes an echo.

**Cycle Map** A specific type of events chain is a cycle map. It is used when the series of events do not produce a final outcome, but instead relate back to the beginning event, such as in **Figure 5.** Therefore, the cycle repeats itself.

To make a cycle map, first decide what event is the beginning event. This is also called the initiating event. Then list the next events in the order that they occur, with the last event relating back to the initiating event. Words can be written between the events that describe what happens from one event to the next. The number of events in a cycle map can vary, but usually contain three or more events.

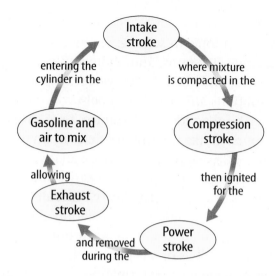

**Figure 5** A cycle map shows events that occur in a cycle.

**Spider Map** A type of concept map that you can use for brainstorming is the spider map. When you have a central idea, you might find that you have a jumble of ideas that relate to it but are not necessarily clearly related to each other. The spider map on sound in **Figure 6** shows that if you write these ideas outside the main concept, then you can begin to separate and group unrelated terms so they become more useful.

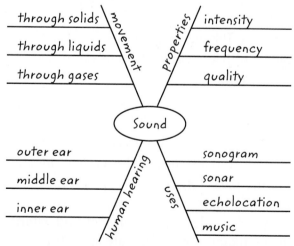

**Figure 6** A spider map allows you to list ideas that relate to a central topic but not necessarily to one another.

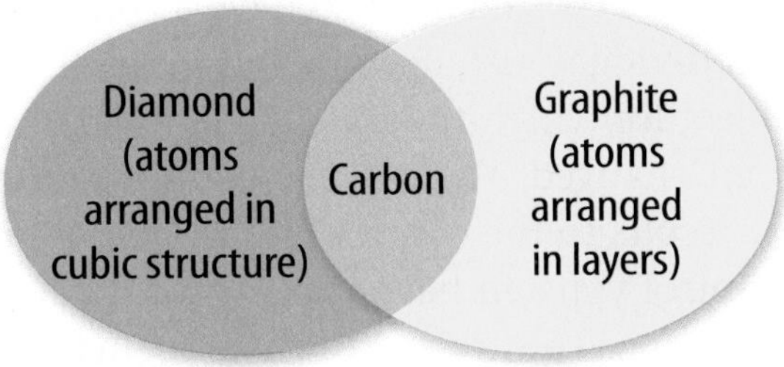

**Figure 7** This Venn diagram compares and contrasts two substances made from carbon.

**Venn Diagram** To illustrate how two subjects compare and contrast you can use a Venn diagram. You can see the characteristics that the subjects have in common and those that they do not, shown in **Figure 7.**

To create a Venn diagram, draw two overlapping ovals that are big enough to write in. List the characteristics unique to one subject in one oval, and the characteristics of the other subject in the other oval. The characteristics in common are listed in the overlapping section.

**Make and Use Tables** One way to organize information so it is easier to understand is to use a table. Tables can contain numbers, words, or both.

To make a table, list the items to be compared in the first column and the characteristics to be compared in the first row. The title should clearly indicate the content of the table, and the column or row heads should be clear. Notice that in **Table 1** the units are included.

| Table 1  Recyclables Collected During Week | | | |
|---|---|---|---|
| **Day of Week** | **Paper (kg)** | **Aluminum (kg)** | **Glass (kg)** |
| Monday | 5.0 | 4.0 | 12.0 |
| Wednesday | 4.0 | 1.0 | 10.0 |
| Friday | 2.5 | 2.0 | 10.0 |

**Make a Model** One way to help you better understand the parts of a structure, the way a process works, or to show things too large or small for viewing is to make a model. For example, an atomic model made of a plastic-ball nucleus and chenille stem electron shells can help you visualize how the parts of an atom relate to each other. Other types of models can be devised on a computer or represented by equations.

## Form a Hypothesis

A possible explanation based on previous knowledge and observations is called a hypothesis. After researching gasoline types and recalling previous experiences in your family's car you form a hypothesis—our car runs more efficiently because we use premium gasoline. To be valid, a hypothesis has to be something you can test by using an investigation.

**Predict** When you apply a hypothesis to a specific situation, you predict something about that situation. A prediction makes a statement in advance, based on prior observation, experience, or scientific reasoning. People use predictions to make everyday decisions. Scientists test predictions by performing investigations. Based on previous observations and experiences, you might form a prediction that cars are more efficient with premium gasoline. The prediction can be tested in an investigation.

**Design an Experiment** A scientist needs to make many decisions before beginning an investigation. Some of these include: how to carry out the investigation, what steps to follow, how to record the data, and how the investigation will answer the question. It also is important to address any safety concerns.

SCIENCE SKILL HANDBOOK

MATH SKILL HANDBOOK

FOLDABLES HANDBOOK

REFERENCE HANDBOOK

GLOSSARY/ GLOSARIO

INDEX

SCIENCE SKILL HANDBOOK

MATH SKILL HANDBOOK

FOLDABLES HANDBOOK

REFERENCE HANDBOOK

GLOSSARY/ GLOSARIO

INDEX

## Test the Hypothesis

Now that you have formed your hypothesis, you need to test it. Using an investigation, you will make observations and collect data, or information. This data might either support or not support your hypothesis. Scientists collect and organize data as numbers and descriptions.

**Follow a Procedure** In order to know what materials to use, as well as how and in what order to use them, you must follow a procedure. **Figure 8** shows a procedure you might follow to test your hypothesis.

---

**Procedure**

**Step 1**   Use regular gasoline for two weeks.

**Step 2**   Record the number of kilometers between fill-ups and the amount of gasoline used.

**Step 3**   Switch to premium gasoline for two weeks.

**Step 4**   Record the number of kilometers between fill-ups and the amount of gasoline used.

---

**Figure 8**  A procedure tells you what to do step-by-step.

**Identify and Manipulate Variables and Controls** In any experiment, it is important to keep everything the same except for the item you are testing. The one factor you change is called the independent variable. The change that results is the dependent variable. Make sure you have only one independent variable, to assure yourself of the cause of the changes you observe in the dependent variable. For example, in your gasoline experiment the type of fuel is the independent variable. The dependent variable is the efficiency.

Many experiments also have a control— an individual instance or experimental subject for which the independent variable is not changed. You can then compare the test results to the control results. To design a control you can have two cars of the same type. The control car uses regular gasoline for four weeks. After you are done with the test, you can compare the experimental results to the control results.

## Collect Data

Whether you are carrying out an investigation or a short observational experiment, you will collect data, as shown in **Figure 9.** Scientists collect data as numbers and descriptions and organize them in specific ways.

**Observe** Scientists observe items and events, then record what they see. When they use only words to describe an observation, it is called qualitative data. Scientists' observations also can describe how much there is of something. These observations use numbers, as well as words, in the description and are called quantitative data. For example, if a sample of the element gold is described as being "shiny and very dense" the data are qualitative. Quantitative data on this sample of gold might include "a mass of 30 g and a density of 19.3 $g/cm^3$."

**Figure 9**  Collecting data is one way to gather information directly.

**Figure 10** Record data neatly and clearly so it is easy to understand.

When you make observations you should examine the entire object or situation first, and then look carefully for details. It is important to record observations accurately and completely. Always record your notes immediately as you make them, so you do not miss details or make a mistake when recording results from memory. Never put unidentified observations on scraps of paper. Instead they should be recorded in a notebook, like the one in **Figure 10.** Write your data neatly so you can easily read it later. At each point in the experiment, record your observations and label them. That way, you will not have to determine what the figures mean when you look at your notes later. Set up any tables that you will need to use ahead of time, so you can record any observations right away. Remember to avoid bias when collecting data by not including personal thoughts when you record observations. Record only what you observe.

**Estimate** Scientific work also involves estimating. To estimate is to make a judgment about the size or the number of something without measuring or counting. This is important when the number or size of an object or population is too large or too difficult to accurately count or measure.

**Sample** Scientists may use a sample or a portion of the total number as a type of estimation. To sample is to take a small, representative portion of the objects or organisms of a population for research. By making careful observations or manipulating variables within that portion of the group, information is discovered and conclusions are drawn that might apply to the whole population. A poorly chosen sample can be unrepresentative of the whole. If you were trying to determine the rainfall in an area, it would not be best to take a rainfall sample from under a tree.

**Measure** You use measurements every day. Scientists also take measurements when collecting data. When taking measurements, it is important to know how to use measuring tools properly. Accuracy also is important.

**Length** To measure length, the distance between two points, scientists use meters. Smaller measurements might be measured in centimeters or millimeters.

Length is measured using a metric ruler or meterstick. When using a metric ruler, line up the 0-cm mark with the end of the object being measured and read the number of the unit where the object ends. Look at the metric ruler shown in **Figure 11.** The centimeter lines are the long, numbered lines, and the shorter lines are millimeter lines. In this instance, the length would be 4.50 cm.

**Figure 11** This metric ruler has centimeter and millimeter divisions.

SCIENCE SKILL HANDBOOK

MATH SKILL HANDBOOK

FOLDABLES HANDBOOK

REFERENCE HANDBOOK

GLOSSARY/ GLOSARIO

INDEX

SCIENCE SKILL HANDBOOK

MATH SKILL HANDBOOK

FOLDABLES HANDBOOK

REFERENCE HANDBOOK

GLOSSARY/ GLOSARIO

INDEX

**Mass** The SI unit for mass is the kilogram (kg). Scientists can measure mass using units formed by adding metric prefixes to the unit gram (g), such as milligram (mg). To measure mass, you might use a triple-beam balance similar to the one shown in **Figure 12.** The balance has a pan on one side and a set of beams on the other side. Each beam has a rider that slides on the beam.

When using a triple-beam balance, place an object on the pan. Slide the largest rider along its beam until the pointer drops below zero. Then move it back one notch. Repeat the process for each rider proceeding from the larger to smaller until the pointer swings an equal distance above and below the zero point. Sum the masses on each beam to find the mass of the object. Move all riders back to zero when finished.

Instead of putting materials directly on the balance, scientists often take a tare of a container. A tare is the mass of a container into which objects or substances are placed for measuring their masses. To find the mass of objects or substances, find the mass of a clean container. Remove the container from the pan, and place the object or substances in the container. Find the mass of the container with the materials in it. Subtract the mass of the empty container from the mass of the filled container to find the mass of the materials you are using.

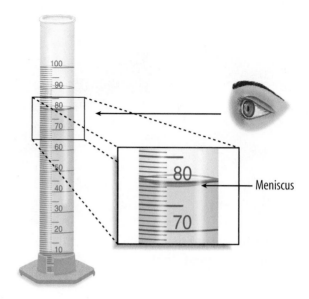

**Figure 13** Graduated cylinders measure liquid volume.

**Liquid Volume** To measure liquids, the unit used is the liter. When a smaller unit is needed, scientists might use a milliliter. Because a milliliter takes up the volume of a cube measuring 1 cm on each side it also can be called a cubic centimeter ($cm^3 = cm \times cm \times cm$).

You can use beakers and graduated cylinders to measure liquid volume. A graduated cylinder, shown in **Figure 13,** is marked from bottom to top in milliliters. In lab, you might use a 10-mL graduated cylinder or a 100-mL graduated cylinder. When measuring liquids, notice that the liquid has a curved surface. Look at the surface at eye level, and measure the bottom of the curve. This is called the meniscus. The graduated cylinder in **Figure 13** contains 79.0 mL, or 79.0 $cm^3$, of a liquid.

**Temperature** Scientists often measure temperature using the Celsius scale. Pure water has a freezing point of 0°C and boiling point of 100°C. The unit of measurement is degrees Celsius. Two other scales often used are the Fahrenheit and Kelvin scales.

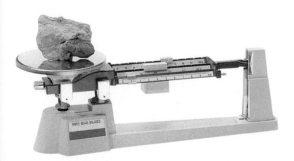

**Figure 12** A triple-beam balance is used to determine the mass of an object.

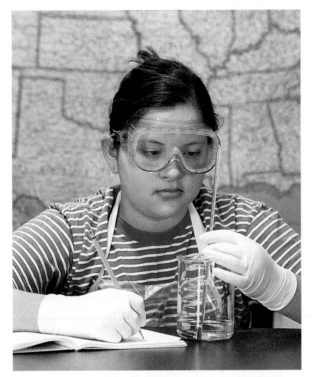

**Figure 14** A thermometer measures the temperature of an object.

Scientists use a thermometer to measure temperature. Most thermometers in a laboratory are glass tubes with a bulb at the bottom end containing a liquid such as colored alcohol. The liquid rises or falls with a change in temperature. To read a glass thermometer like the thermometer in **Figure 14,** rotate it slowly until a red line appears. Read the temperature where the red line ends.

**Form Operational Definitions** An operational definition defines an object by how it functions, works, or behaves. For example, when you are playing hide and seek and a tree is home base, you have created an operational definition for a tree.

Objects can have more than one operational definition. For example, a ruler can be defined as a tool that measures the length of an object (how it is used). It can also be a tool with a series of marks used as a standard when measuring (how it works).

## Analyze the Data

To determine the meaning of your observations and investigation results, you will need to look for patterns in the data. Then you must think critically to determine what the data mean. Scientists use several approaches when they analyze the data they have collected and recorded. Each approach is useful for identifying specific patterns.

**Interpret Data** The word *interpret* means "to explain the meaning of something." When analyzing data from an experiment, try to find out what the data show. Identify the control group and the test group to see whether changes in the independent variable have had an effect. Look for differences in the dependent variable between the control and test groups.

**Classify** Sorting objects or events into groups based on common features is called classifying. When classifying, first observe the objects or events to be classified. Then select one feature that is shared by some members in the group, but not by all. Place those members that share that feature in a subgroup. You can classify members into smaller and smaller subgroups based on characteristics. Remember that when you classify, you are grouping objects or events for a purpose. Keep your purpose in mind as you select the features to form groups and subgroups.

**Compare and Contrast** Observations can be analyzed by noting the similarities and differences between two or more objects or events that you observe. When you look at objects or events to see how they are similar, you are comparing them. Contrasting is looking for differences in objects or events.

SCIENCE SKILL HANDBOOK

MATH SKILL HANDBOOK

FOLDABLES HANDBOOK

REFERENCE HANDBOOK

GLOSSARY/GLOSARIO

INDEX

**Recognize Cause and Effect**  A cause is a reason for an action or condition. The effect is that action or condition. When two events happen together, it is not necessarily true that one event caused the other. Scientists must design a controlled investigation to recognize the exact cause and effect.

## Draw Conclusions

When scientists have analyzed the data they collected, they proceed to draw conclusions about the data. These conclusions are sometimes stated in words similar to the hypothesis that you formed earlier. They may confirm a hypothesis, or lead you to a new hypothesis.

**Infer**  Scientists often make inferences based on their observations. An inference is an attempt to explain observations or to indicate a cause. An inference is not a fact, but a logical conclusion that needs further investigation. For example, you may infer that a fire has caused smoke. Until you investigate, however, you do not know for sure.

**Apply**  When you draw a conclusion, you must apply those conclusions to determine whether the data supports the hypothesis. If your data do not support your hypothesis, it does not mean that the hypothesis is wrong. It means only that the result of the investigation did not support the hypothesis. Maybe the experiment needs to be redesigned, or some of the initial observations on which the hypothesis was based were incomplete or biased. Perhaps more observation or research is needed to refine your hypothesis. A successful investigation does not always come out the way you originally predicted.

**Avoid Bias**  Sometimes a scientific investigation involves making judgments. When you make a judgment, you form an opinion. It is important to be honest and not to allow any expectations of results to bias your judgments. This is important throughout the entire investigation, from researching to collecting data to drawing conclusions.

## Communicate

The communication of ideas is an important part of the work of scientists. A discovery that is not reported will not advance the scientific community's understanding or knowledge. Communication among scientists also is important as a way of improving their investigations.

Scientists communicate in many ways, from writing articles in journals and magazines that explain their investigations and experiments, to announcing important discoveries on television and radio. Scientists also share ideas with colleagues on the Internet or present them as lectures, like the student is doing in **Figure 15.**

**Figure 15**  A student communicates to his peers about his investigation.

These safety symbols are used in laboratory and field investigations in this book to indicate possible hazards. Learn the meaning of each symbol and refer to this page often. *Remember to wash your hands thoroughly after completing lab procedures.*

## PROTECTIVE EQUIPMENT   Do not begin any lab without the proper protection equipment.

| | | |
|---|---|---|
|  **GOGGLES** | Proper eye protection must be worn when performing or observing science activities that involve items or conditions as listed below. | |
|  **APRON** | Wear an approved apron when using substances that could stain, wet, or destroy cloth. | |
|  **SOAP** | Wash hands with soap and water before removing goggles and after all lab activities. | |
|  **GLOVES** | Wear gloves when working with biological materials, chemicals, animals, or materials that can stain or irritate hands. | |

## LABORATORY HAZARDS

| Symbols | Potential Hazards | Precaution | Response |
|---|---|---|---|
| **DISPOSAL** | contamination of classroom or environment due to improper disposal of materials such as chemicals and live specimens | • DO NOT dispose of hazardous materials in the sink or trash can.<br>• Dispose of wastes as directed by your teacher. | • If hazardous materials are disposed of improperly, notify your teacher immediately. |
| **EXTREME TEMPERATURE** | skin burns due to extremely hot or cold materials such as hot glass, liquids, or metals; liquid nitrogen; dry ice | • Use proper protective equipment, such as hot mitts and/or tongs, when handling objects with extreme temperatures. | • If injury occurs, notify your teacher immediately. |
| **SHARP OBJECTS** | punctures or cuts from sharp objects such as razor blades, pins, scalpels, and broken glass | • Handle glassware carefully to avoid breakage.<br>• Walk with sharp objects pointed downward, away from you and others. | • If broken glass or injury occurs, notify your teacher immediately. |
| **ELECTRICAL** | electric shock or skin burn due to improper grounding, short circuits, liquid spills, or exposed wires | • Check condition of wires and apparatus for fraying or uninsulated wires, and broken or cracked equipment.<br>• Use only GFCI-protected outlets | • DO NOT attempt to fix electrical problems. Notify your teacher immediately. |
| **CHEMICAL** | skin irritation or burns, breathing difficulty, and/or poisoning due to touching, swallowing, or inhalation of chemicals such as acids, bases, bleach, metal compounds, iodine, poinsettias, pollen, ammonia, acetone, nail polish remover, heated chemicals, mothballs, and any other chemicals labeled or known to be dangerous | • Wear proper protective equipment such as goggles, apron, and gloves when using chemicals.<br>• Ensure proper room ventilation or use a fume hood when using materials that produce fumes.<br>• NEVER smell fumes directly.<br>• NEVER taste or eat any material in the laboratory. | • If contact occurs, immediately flush affected area with water and notify your teacher.<br>• If a spill occurs, leave the area immediately and notify your teacher. |
| **FLAMMABLE** | unexpected fire due to liquids or gases that ignite easily such as rubbing alcohol | • Avoid open flames, sparks, or heat when flammable liquids are present. | • If a fire occurs, leave the area immediately and notify your teacher. |
| **OPEN FLAME** | burns or fire due to open flame from matches, Bunsen burners, or burning materials | • Tie back loose hair and clothing.<br>• Keep flame away from all materials.<br>• Follow teacher instructions when lighting and extinguishing flames.<br>• Use proper protection, such as hot mitts or tongs, when handling hot objects. | • If a fire occurs, leave the area immediately and notify your teacher. |
| **ANIMAL SAFETY** | injury to or from laboratory animals | • Wear proper protective equipment such as gloves, apron, and goggles when working with animals.<br>• Wash hands after handling animals. | • If injury occurs, notify your teacher immediately. |
| **BIOLOGICAL** | infection or adverse reaction due to contact with organisms such as bacteria, fungi, and biological materials such as blood, animal or plant materials | • Wear proper protective equipment such as gloves, goggles, and apron when working with biological materials.<br>• Avoid skin contact with an organism or any part of the organism.<br>• Wash hands after handling organisms. | • If contact occurs, wash the affected area and notify your teacher immediately. |
| **FUME** | breathing difficulties from inhalation of fumes from substances such as ammonia, acetone, nail polish remover, heated chemicals, and mothballs | • Wear goggles, apron, and gloves.<br>• Ensure proper room ventilation or use a fume hood when using substances that produce fumes.<br>• NEVER smell fumes directly. | • If a spill occurs, leave area and notify your teacher immediately. |
| **IRRITANT** | irritation of skin, mucous membranes, or respiratory tract due to materials such as acids, bases, bleach, pollen, mothballs, steel wool, and potassium permanganate | • Wear goggles, apron, and gloves.<br>• Wear a dust mask to protect against fine particles. | • If skin contact occurs, immediately flush the affected area with water and notify your teacher. |
| **RADIOACTIVE** | excessive exposure from alpha, beta, and gamma particles | • Remove gloves and wash hands with soap and water before removing remainder of protective equipment. | • If cracks or holes are found in the container, notify your teacher immediately. |

SCIENCE SKILL HANDBOOK

MATH SKILL HANDBOOK

FOLDABLES HANDBOOK

REFERENCE HANDBOOK

GLOSSARY/ GLOSARIO

INDEX

# Safety in the Science Laboratory

## Introduction to Science Safety

The science laboratory is a safe place to work if you follow standard safety procedures. Being responsible for your own safety helps to make the entire laboratory a safer place for everyone. When performing any lab, read and apply the caution statements and safety symbol listed at the beginning of the lab.

## General Safety Rules

1. Complete the *Lab Safety Form* or other safety contract BEFORE starting any science lab.

2. Study the procedure. Ask your teacher any questions. Be sure you understand safety symbols shown on the page.

3. Notify your teacher about allergies or other health conditions that can affect your participation in a lab.

4. Learn and follow use and safety procedures for your equipment. If unsure, ask your teacher.

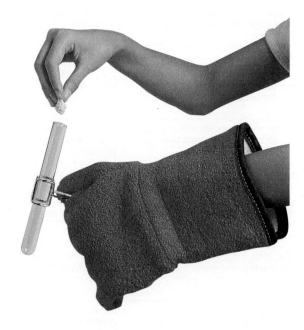

5. Never eat, drink, chew gum, apply cosmetics, or do any personal grooming in the lab. Never use lab glassware as food or drink containers. Keep your hands away from your face and mouth.

6. Know the location and proper use of the safety shower, eye wash, fire blanket, and fire alarm.

## Prevent Accidents

1. Use the safety equipment provided to you. Goggles and a safety apron should be worn during investigations.

2. Do NOT use hair spray, mousse, or other flammable hair products. Tie back long hair and tie down loose clothing.

3. Do NOT wear sandals or other open-toed shoes in the lab.

4. Remove jewelry on hands and wrists. Loose jewelry, such as chains and long necklaces, should be removed to prevent them from getting caught in equipment.

5. Do not taste any substances or draw any material into a tube with your mouth.

6. Proper behavior is expected in the lab. Practical jokes and fooling around can lead to accidents and injury.

7. Keep your work area uncluttered.

## Laboratory Work

1. Collect and carry all equipment and materials to your work area before beginning a lab.

2. Remain in your own work area unless given permission by your teacher to leave it.

SCIENCE SKILL HANDBOOK

MATH SKILL HANDBOOK

FOLDABLES HANDBOOK

REFERENCE HANDBOOK

GLOSSARY/ GLOSARIO

INDEX

3. Always slant test tubes away from your-self and others when heating them, adding substances to them, or rinsing them.

4. If instructed to smell a substance in a container, hold the container a short distance away and fan vapors toward your nose.

5. Do NOT substitute other chemicals/substances for those in the materials list unless instructed to do so by your teacher.

6. Do NOT take any materials or chemicals outside of the laboratory.

7. Stay out of storage areas unless instructed to be there and supervised by your teacher.

## Laboratory Cleanup

1. Turn off all burners, water, and gas, and disconnect all electrical devices.

2. Clean all pieces of equipment and return all materials to their proper places.

3. Dispose of chemicals and other materials as directed by your teacher. Place broken glass and solid substances in the proper containers. Never discard materials in the sink.

4. Clean your work area.

5. Wash your hands with soap and water thoroughly BEFORE removing your goggles.

## Emergencies

1. Report any fire, electrical shock, glass-ware breakage, spill, or injury, no matter how small, to your teacher immediately. Follow his or her instructions.

2. If your clothing should catch fire, STOP, DROP, and ROLL. If possible, smother it with the fire blanket or get under a safety shower. NEVER RUN.

3. If a fire should occur, turn off all gas and leave the room according to established procedures.

4. In most instances, your teacher will clean up spills. Do NOT attempt to clean up spills unless you are given permission and instructions to do so.

5. If chemicals come into contact with your eyes or skin, notify your teacher immediately. Use the eyewash, or flush your skin or eyes with large quantities of water.

6. The fire extinguisher and first-aid kit should only be used by your teacher unless it is an extreme emergency and you have been given permission.

7. If someone is injured or becomes ill, only a professional medical provider or someone certified in first aid should perform first-aid procedures.

SCIENCE SKILL HANDBOOK

MATH SKILL HANDBOOK

FOLDABLES HANDBOOK

REFERENCE HANDBOOK

GLOSSARY/ GLOSARIO

INDEX

## Use Fractions

A fraction compares a part to a whole. In the fraction $\frac{2}{3}$, the 2 represents the part and is the numerator. The 3 represents the whole and is the denominator.

**Reduce Fractions** To reduce a fraction, you must find the largest factor that is common to both the numerator and the denominator, the greatest common factor (GCF). Divide both numbers by the GCF. The fraction has then been reduced, or it is in its simplest form.

### Example

Twelve of the 20 chemicals in the science lab are in powder form. What fraction of the chemicals used in the lab are in powder form?

**Step 1**  Write the fraction.

$$\frac{part}{whole} = \frac{12}{20}$$

**Step 2**  To find the GCF of the numerator and denominator, list all of the factors of each number.

Factors of 12: 1, 2, 3, 4, 6, 12 (the numbers that divide evenly into 12)

Factors of 20: 1, 2, 4, 5, 10, 20 (the numbers that divide evenly into 20)

**Step 3**  List the common factors.

1, 2, 4

**Step 4**  Choose the greatest factor in the list. The GCF of 12 and 20 is 4.

**Step 5**  Divide the numerator and denominator by the GCF.

$$\frac{12 \div 4}{20 \div 4} = \frac{3}{5}$$

In the lab, $\frac{3}{5}$ of the chemicals are in powder form.

**Practice Problem** At an amusement park, 66 of 90 rides have a height restriction. What fraction of the rides, in its simplest form, has a height restriction?

**Add and Subtract Fractions with Like Denominators** To add or subtract fractions with the same denominator, add or subtract the numerators and write the sum or difference over the denominator. After finding the sum or difference, find the simplest form for your fraction.

### Example 1

In the forest outside your house, $\frac{1}{8}$ of the animals are rabbits, $\frac{3}{8}$ are squirrels, and the remainder are birds and insects. How many are mammals?

**Step 1**  Add the numerators.

$$\frac{1}{8} + \frac{3}{8} = \frac{(1 + 3)}{8} = \frac{4}{8}$$

**Step 2**  Find the GCF.

$$\frac{4}{8} \text{ (GCF, 4)}$$

**Step 3**  Divide the numerator and denominator by the GCF.

$$\frac{4 \div 4}{8 \div 4} = \frac{1}{2}$$

$\frac{1}{2}$ of the animals are mammals.

### Example 2

If $\frac{7}{16}$ of the Earth is covered by freshwater, and $\frac{1}{16}$ of that is in glaciers, how much freshwater is not frozen?

**Step 1**  Subtract the numerators.

$$\frac{7}{16} - \frac{1}{16} = \frac{(7 - 1)}{16} = \frac{6}{16}$$

**Step 2**  Find the GCF.

$$\frac{6}{16} \text{ (GCF, 2)}$$

**Step 3**  Divide the numerator and denominator by the GCF.

$$\frac{6 \div 2}{16 \div 2} = \frac{3}{8}$$

$\frac{3}{8}$ of the freshwater is not frozen.

**Practice Problem** A bicycle rider is riding at a rate of 15 km/h for $\frac{4}{9}$ of his ride, 10 km/h for $\frac{2}{9}$ of his ride, and 8 km/h for the remainder of the ride. How much of his ride is he riding at a rate greater than 8 km/h?

SCIENCE SKILL HANDBOOK

MATH SKILL HANDBOOK

REFERENCE HANDBOOK

GLOSSARY/ GLOSARIO

INDEX

**Add and Subtract Fractions with Unlike Denominators** To add or subtract fractions with unlike denominators, first find the least common denominator (LCD). This is the smallest number that is a common multiple of both denominators. Rename each fraction with the LCD, and then add or subtract. Find the simplest form if necessary.

## Example 1

A chemist makes a paste that is $\frac{1}{2}$ table salt (NaCl), $\frac{1}{3}$ sugar ($C_6H_{12}O_6$), and the remainder is water ($H_2O$). How much of the paste is a solid?

**Step 1** Find the LCD of the fractions.

$\frac{1}{2} + \frac{1}{3}$ (LCD, 6)

**Step 2** Rename each numerator and each denominator with the LCD.

**Step 3** Add the numerators.

$\frac{3}{6} + \frac{2}{6} = \frac{(3+2)}{6} = \frac{5}{6}$

$\frac{5}{6}$ of the paste is a solid.

## Example 2

The average precipitation in Grand Junction, CO, is $\frac{7}{10}$ inch in November, and $\frac{3}{5}$ inch in December. What is the total average precipitation?

**Step 1** Find the LCD of the fractions.

$\frac{7}{10} + \frac{3}{5}$ (LCD, 10)

**Step 2** Rename each numerator and each denominator with the LCD.

**Step 3** Add the numerators.

$\frac{7}{10} + \frac{6}{10} = \frac{(7+6)}{10} = \frac{13}{10}$

$\frac{13}{10}$ inches total precipitation, or $1\frac{3}{10}$ inches.

**Practice Problem** On an electric bill, about $\frac{1}{8}$ of the energy is from solar energy and about $\frac{1}{10}$ is from wind power. How much of the total bill is from solar energy and wind power combined?

## Example 3

In your body, $\frac{7}{10}$ of your muscle contractions are involuntary (cardiac and smooth muscle tissue). Smooth muscle makes $\frac{3}{15}$ of your muscle contractions. How many of your muscle contractions are made by cardiac muscle?

**Step 1** Find the LCD of the fractions.

$\frac{7}{10} - \frac{3}{15}$ (LCD, 30)

**Step 2** Rename each numerator and each denominator with the LCD.

$\frac{7 \times 3}{10 \times 3} = \frac{21}{30}$

$\frac{3 \times 2}{15 \times 2} = \frac{6}{30}$

**Step 3** Subtract the numerators.

$\frac{21}{30} - \frac{6}{30} = \frac{(21-6)}{30} = \frac{15}{30}$

**Step 4** Find the GCF.

$\frac{15}{30}$ (GCF, 15)

$\frac{1}{2}$

$\frac{1}{2}$ of all muscle contractions are cardiac muscle.

## Example 4

Tony wants to make cookies that call for $\frac{3}{4}$ of a cup of flour, but he only has $\frac{1}{3}$ of a cup. How much more flour does he need?

**Step 1** Find the LCD of the fractions.

$\frac{3}{4} - \frac{1}{3}$ (LCD, 12)

**Step 2** Rename each numerator and each denominator with the LCD.

$\frac{3 \times 3}{4 \times 3} = \frac{9}{12}$

$\frac{1 \times 4}{3 \times 4} = \frac{4}{12}$

**Step 3** Subtract the numerators.

$\frac{9}{12} - \frac{4}{12} = \frac{(9-4)}{12} = \frac{5}{12}$

$\frac{5}{12}$ of a cup of flour

**Practice Problem** Using the information provided to you in Example 3 above, determine how many muscle contractions are voluntary (skeletal muscle).

**Multiply Fractions** To multiply with fractions, multiply the numerators and multiply the denominators. Find the simplest form if necessary.

### Example

Multiply $\frac{3}{5}$ by $\frac{1}{3}$.

**Step 1** Multiply the numerators and denominators.

$$\frac{3}{5} \times \frac{1}{3} = \frac{(3 \times 1)}{(5 \times 3)} \frac{3}{15}$$

**Step 2** Find the GCF.

$$\frac{3}{15} \text{ (GCF, 3)}$$

**Step 3** Divide the numerator and denominator by the GCF.

$$\frac{3 \div 3}{15 \div 3} = \frac{1}{5}$$

$\frac{3}{5}$ multiplied by $\frac{1}{3}$ is $\frac{1}{5}$.

**Practice Problem** Multiply $\frac{3}{14}$ by $\frac{5}{16}$.

**Find a Reciprocal** Two numbers whose product is 1 are called multiplicative inverses, or reciprocals.

### Example

Find the reciprocal of $\frac{3}{8}$.

**Step 1** Inverse the fraction by putting the denominator on top and the numerator on the bottom.

$$\frac{8}{3}$$

The reciprocal of $\frac{3}{8}$ is $\frac{8}{3}$.

**Practice Problem** Find the reciprocal of $\frac{4}{9}$.

**Divide Fractions** To divide one fraction by another fraction, multiply the dividend by the reciprocal of the divisor. Find the simplest form if necessary.

### Example 1

Divide $\frac{1}{9}$ by $\frac{1}{3}$.

**Step 1** Find the reciprocal of the divisor.

The reciprocal of $\frac{1}{3}$ is $\frac{3}{1}$.

**Step 2** Multiply the dividend by the reciprocal of the divisor.

$$\frac{\frac{1}{9}}{\frac{1}{3}} = \frac{1}{9} \times \frac{3}{1} = \frac{(1 \times 3)}{(9 \times 1)} = \frac{3}{9}$$

**Step 3** Find the GCF.

$$\frac{3}{9} \text{ (GCF, 3)}$$

**Step 4** Divide the numerator and denominator by the GCF.

$$\frac{3 \div 3}{9 \div 3} = \frac{1}{3}$$

$\frac{1}{9}$ divided by $\frac{1}{3}$ is $\frac{1}{3}$.

### Example 2

Divide $\frac{3}{5}$ by $\frac{1}{4}$.

**Step 1** Find the reciprocal of the divisor.

The reciprocal of $\frac{1}{4}$ is $\frac{4}{1}$.

**Step 2** Multiply the dividend by the reciprocal of the divisor.

$$\frac{\frac{3}{5}}{\frac{1}{4}} = \frac{3}{5} \times \frac{4}{1} = \frac{(3 \times 4)}{(5 \times 1)} = \frac{12}{5}$$

$\frac{3}{5}$ divided by $\frac{1}{4}$ is $\frac{12}{5}$ or $2\frac{2}{5}$.

**Practice Problem** Divide $\frac{3}{11}$ by $\frac{7}{10}$.

## Use Ratios

When you compare two numbers by division, you are using a ratio. Ratios can be written 3 to 5, 3:5, or $\frac{3}{5}$. Ratios, like fractions, also can be written in simplest form.

Ratios can represent one type of probability, called odds. This is a ratio that compares the number of ways a certain outcome occurs to the number of possible outcomes. For example, if you flip a coin 100 times, what are the odds that it will come up heads? There are two possible outcomes, heads or tails, so the odds of coming up heads are 50:100. Another way to say this is that 50 out of 100 times the coin will come up heads. In its simplest form, the ratio is 1:2.

### Example 1

A chemical solution contains 40 g of salt and 64 g of baking soda. What is the ratio of salt to baking soda as a fraction in simplest form?

**Step 1**   Write the ratio as a fraction.

$$\frac{\text{salt}}{\text{baking soda}} = \frac{40}{64}$$

**Step 2**   Express the fraction in simplest form. The GCF of 40 and 64 is 8.

$$\frac{40}{64} = \frac{40 \div 8}{64 \div 8} = \frac{5}{8}$$

The ratio of salt to baking soda in the sample is 5:8.

### Example 2

Sean rolls a 6-sided die 6 times. What are the odds that the side with a 3 will show?

**Step 1**   Write the ratio as a fraction.

$$\frac{\text{number of sides with a 3}}{\text{number of possible sides}} = \frac{1}{6}$$

**Step 2**   Multiply by the number of attempts.

$$\frac{1}{6} \times 6 \text{ attempts} = \frac{6}{6} \text{ attempts} = 1 \text{ attempt}$$

1 attempt out of 6 will show a 3.

**Practice Problem**   Two metal rods measure 100 cm and 144 cm in length. What is the ratio of their lengths in simplest form?

## Use Decimals

A fraction with a denominator that is a power of ten can be written as a decimal. For example, 0.27 means $\frac{27}{100}$. The decimal point separates the ones place from the tenths place.

Any fraction can be written as a decimal using division. For example, the fraction $\frac{5}{8}$ can be written as a decimal by dividing 5 by 8. Written as a decimal, it is 0.625.

**Add or Subtract Decimals**   When adding and subtracting decimals, line up the decimal points before carrying out the operation.

### Example 1

Find the sum of 47.68 and 7.80.

**Step 1**   Line up the decimal places when you write the numbers.

$$\begin{array}{r} 47.68 \\ + \ 7.80 \end{array}$$

**Step 2**   Add the decimals.

$$\begin{array}{r} \overset{1\ 1}{47.68} \\ + \ 7.80 \\ \hline 55.48 \end{array}$$

The sum of 47.68 and 7.80 is 55.48.

### Example 2

Find the difference of 42.17 and 15.85.

**Step 1**   Line up the decimal places when you write the number.

$$\begin{array}{r} 42.17 \\ -15.85 \end{array}$$

**Step 2**   Subtract the decimals.

$$\begin{array}{r} \overset{3\,11}{4\cancel{2}.1\overset{1}{7}} \\ -15.85 \\ \hline 26.32 \end{array}$$

The difference of 42.17 and 15.85 is 26.32.

**Practice Problem**   Find the sum of 1.245 and 3.842.

**Multiply Decimals** To multiply decimals, multiply the numbers like numbers without decimal points. Count the decimal places in each factor. The product will have the same number of decimal places as the sum of the decimal places in the factors.

### Example

Multiply 2.4 by 5.9.

**Step 1**   Multiply the factors like two whole numbers.

$24 \times 59 = 1416$

**Step 2**   Find the sum of the number of decimal places in the factors. Each factor has one decimal place, for a sum of two decimal places.

**Step 3**   The product will have two decimal places.

14.16

The product of 2.4 and 5.9 is 14.16.

**Practice Problem**   Multiply 4.6 by 2.2.

**Divide Decimals** When dividing decimals, change the divisor to a whole number. To do this, multiply both the divisor and the dividend by the same power of ten. Then place the decimal point in the quotient directly above the decimal point in the dividend. Then divide as you do with whole numbers.

### Example

Divide 8.84 by 3.4.

**Step 1**   Multiply both factors by 10.

$3.4 \times 10 = 34, 8.84 \times 10 = 88.4$

**Step 2**   Divide 88.4 by 34.

$$\begin{array}{r} 2.6 \\ 34\overline{)88.4} \\ -68\phantom{.} \\ \hline 204 \\ -204 \\ \hline 0 \end{array}$$

8.84 divided by 3.4 is 2.6.

**Practice Problem**   Divide 75.6 by 3.6.

## Use Proportions

An equation that shows that two ratios are equivalent is a proportion. The ratios $\frac{2}{4}$ and $\frac{5}{10}$ are equivalent, so they can be written as $\frac{2}{4} = \frac{5}{10}$. This equation is a proportion.

When two ratios form a proportion, the cross products are equal. To find the cross products in the proportion $\frac{2}{4} = \frac{5}{10}$, multiply the 2 and the 10, and the 4 and the 5. Therefore $2 \times 10 = 4 \times 5$, or $20 = 20$.

Because you know that both ratios are equal, you can use cross products to find a missing term in a proportion. This is known as solving the proportion.

### Example

The heights of a tree and a pole are proportional to the lengths of their shadows. The tree casts a shadow of 24 m when a 6-m pole casts a shadow of 4 m. What is the height of the tree?

**Step 1**   Write a proportion.

$$\frac{\text{height of tree}}{\text{height of pole}} = \frac{\text{length of tree's shadow}}{\text{length of pole's shadow}}$$

**Step 2**   Substitute the known values into the proportion. Let $h$ represent the unknown value, the height of the tree.

$$\frac{h}{6} \times \frac{24}{4}$$

**Step 3**   Find the cross products.

$$h \times 4 = 6 \times 24$$

**Step 4**   Simplify the equation.

$$4h \times 144$$

**Step 5**   Divide each side by 4.

$$\frac{4h}{4} \times \frac{144}{4}$$

$$h = 36$$

The height of the tree is 36 m.

**Practice Problem**   The ratios of the weights of two objects on the Moon and on Earth are in proportion. A rock weighing 3 N on the Moon weighs 18 N on Earth. How much would a rock that weighs 5 N on the Moon weigh on Earth?

## Use Percentages

The word *percent* means "out of one hundred." It is a ratio that compares a number to 100. Suppose you read that 77 percent of Earth's surface is covered by water. That is the same as reading that the fraction of Earth's surface covered by water is $\frac{77}{100}$. To express a fraction as a percent, first find the equivalent decimal for the fraction. Then, multiply the decimal by 100 and add the percent symbol.

### Example 1

Express $\frac{13}{20}$ as a percent.

**Step 1** Find the equivalent decimal for the fraction.

$$\begin{array}{r} 0.65 \\ 20\overline{)13.00} \\ \underline{12\,0}\phantom{0} \\ 1\,00 \\ \underline{1\,00} \\ 0 \end{array}$$

**Step 2** Rewrite the fraction $\frac{13}{20}$ as 0.65.

**Step 3** Multiply 0.65 by 100 and add the % symbol.

$$0.65 \times 100 = 65 = 65\%$$

So, $\frac{13}{20} = 65\%$.

This also can be solved as a proportion.

### Example 2

Express $\frac{13}{20}$ as a percent.

**Step 1** Write a proportion.

$$\frac{13}{20} = \frac{x}{100}$$

**Step 2** Find the cross products.

$$1300 = 20x$$

**Step 3** Divide each side by 20.

$$\frac{1300}{20} = \frac{20x}{20}$$

$$65\% = x$$

**Practice Problem** In one year, 73 of 365 days were rainy in one city. What percent of the days in that city were rainy?

## Solve One-Step Equations

A statement that two expressions are equal is an equation. For example, $A = B$ is an equation that states that $A$ is equal to $B$.

An equation is solved when a variable is replaced with a value that makes both sides of the equation equal. To make both sides equal the inverse operation is used. Addition and subtraction are inverses, and multiplication and division are inverses.

### Example 1

Solve the equation $x - 10 = 35$.

**Step 1** Find the solution by adding 10 to each side of the equation.

$$x - 10 = 35$$
$$x - 10 + 10 = 35 - 10$$
$$x = 45$$

**Step 2** Check the solution.

$$x - 10 = 35$$
$$45 - 10 = 35$$
$$35 = 35$$

Both sides of the equation are equal, so $x = 45$.

### Example 2

In the formula $a = bc$, find the value of $c$ if $a = 20$ and $b = 2$.

**Step 1** Rearrange the formula so the unknown value is by itself on one side of the equation by dividing both sides by $b$.

$$a = bc$$
$$\frac{a}{b} = \frac{bc}{b}$$
$$\frac{a}{b} = c$$

**Step 2** Replace the variables $a$ and $b$ with the values that are given.

$$\frac{a}{b} = c$$
$$\frac{20}{2} = c$$
$$10 = c$$

**Step 3** Check the solution.

$$a = bc$$
$$20 = 2 \times 10$$
$$20 = 20$$

Both sides of the equation are equal, so $c = 10$ is the solution when $a = 20$ and $b = 2$.

**Practice Problem** In the formula $h = gd$, find the value of $d$ if $g = 12.3$ and $h = 17.4$.

## Use Statistics

The branch of mathematics that deals with collecting, analyzing, and presenting data is statistics. In statistics, there are three common ways to summarize data with a single number—the mean, the median, and the mode.

The **mean** of a set of data is the arithmetic average. It is found by adding the numbers in the data set and dividing by the number of items in the set.

The **median** is the middle number in a set of data when the data are arranged in numerical order. If there were an even number of data points, the median would be the mean of the two middle numbers.

The **mode** of a set of data is the number or item that appears most often.

Another number that often is used to describe a set of data is the range. The **range** is the difference between the largest number and the smallest number in a set of data.

### Example

The speeds (in m/s) for a race car during five different time trials are 39, 37, 44, 36, and 44.

**To find the mean:**

**Step 1** Find the sum of the numbers.

$$39 + 37 + 44 + 36 + 44 = 200$$

**Step 2** Divide the sum by the number of items, which is 5.

$$200 \div 5 = 40$$

The mean is 40 m/s.

**To find the median:**

**Step 1** Arrange the measures from least to greatest.

36, 37, 39, 44, 44

**Step 2** Determine the middle measure.

36, 37, <u>39</u>, 44, 44

The median is 39 m/s.

**To find the mode:**

**Step 1** Group the numbers that are the same together.

44, 44, 36, 37, 39

**Step 2** Determine the number that occurs most in the set.

<u>44, 44</u>, 36, 37, 39

The mode is 44 m/s.

**To find the range:**

**Step 1** Arrange the measures from greatest to least.

44, 44, 39, 37, 36

**Step 2** Determine the greatest and least measures in the set.

<u>44</u>, 44, 39, 37, <u>36</u>

**Step 3** Find the difference between the greatest and least measures.

$$44 - 36 = 8$$

The range is 8 m/s.

**Practice Problem** Find the mean, median, mode, and range for the data set 8, 4, 12, 8, 11, 14, 16.

A **frequency table** shows how many times each piece of data occurs, usually in a survey. **Table 1** below shows the results of a student survey on favorite color.

| Table 1 Student Color Choice | | |
| --- | --- | --- |
| Color | Tally | Frequency |
| red | IIII | 4 |
| blue | ʬ | 5 |
| black | II | 2 |
| green | III | 3 |
| purple | ʬ II | 7 |
| yellow | ʬ I | 6 |

Based on the frequency table data, which color is the favorite?

## Use Geometry

The branch of mathematics that deals with the measurement, properties, and relationships of points, lines, angles, surfaces, and solids is called geometry.

**Perimeter** The **perimeter** ($P$) is the distance around a geometric figure. To find the perimeter of a rectangle, add the length and width and multiply that sum by two, or $2(l + w)$. To find perimeters of irregular figures, add the length of the sides.

### Example 1

Find the perimeter of a rectangle that is 3 m long and 5 m wide.

**Step 1** You know that the perimeter is 2 times the sum of the width and length.

$$P = 2(3 \text{ m} + 5 \text{ m})$$

**Step 2** Find the sum of the width and length.

$$P = 2(8 \text{ m})$$

**Step 3** Multiply by 2.

$$P = 16 \text{ m}$$

The perimeter is 16 m.

### Example 2

Find the perimeter of a shape with sides measuring 2 cm, 5 cm, 6 cm, 3 cm.

**Step 1** You know that the perimeter is the sum of all the sides.

$$P = 2 + 5 + 6 + 3$$

**Step 2** Find the sum of the sides.

$$P = 2 + 5 + 6 + 3$$
$$P = 16$$

The perimeter is 16 cm.

**Practice Problem** Find the perimeter of a rectangle with a length of 18 m and a width of 7 m.

**Practice Problem** Find the perimeter of a triangle measuring 1.6 cm by 2.4 cm by 2.4 cm.

**Area of a Rectangle** The **area** ($A$) is the number of square units needed to cover a surface. To find the area of a rectangle, multiply the length times the width, or $l \times w$. When finding area, the units also are multiplied. Area is given in square units.

### Example

Find the area of a rectangle with a length of 1 cm and a width of 10 cm.

**Step 1** You know that the area is the length multiplied by the width.

$$A = (1 \text{ cm} \times 10 \text{ cm})$$

**Step 2** Multiply the length by the width. Also multiply the units.

$$A = 10 \text{ cm}^2$$

The area is 10 cm².

**Practice Problem** Find the area of a square whose sides measure 4 m.

**Area of a Triangle** To find the area of a triangle, use the formula:

$$A = \frac{1}{2}(\text{base} \times \text{height})$$

The base of a triangle can be any of its sides. The height is the perpendicular distance from a base to the opposite endpoint, or vertex.

### Example

Find the area of a triangle with a base of 18 m and a height of 7 m.

**Step 1** You know that the area is $\frac{1}{2}$ the base times the height.

$$A = \frac{1}{2}(18 \text{ m} \times 7 \text{ m})$$

**Step 2** Multiply $\frac{1}{2}$ by the product of $18 \times 7$. Multiply the units.

$$A = \frac{1}{2}(126 \text{ m}^2)$$
$$A = 63 \text{ m}^2$$

The area is 63 m².

**Practice Problem** Find the area of a triangle with a base of 27 cm and a height of 17 cm.

**Circumference of a Circle** The **diameter** (*d*) of a circle is the distance across the circle through its center, and the **radius** (r) is the distance from the center to any point on the circle. The radius is half of the diameter. The distance around the circle is called the **circumference** (C). The formula for finding the circumference is:

$$C = 2\pi r \text{ or } C = \pi d$$

The circumference divided by the diameter is always equal to 3.1415926... This nonterminating and nonrepeating number is represented by the Greek letter $\pi$ (pi). An approximation often used for $\pi$ is 3.14.

### Example 1

Find the circumference of a circle with a radius of 3 m.

**Step 1** You know the formula for the circumference is 2 times the radius times $\pi$.

$$C = 2\pi(3)$$

**Step 2** Multiply 2 times the radius.

$$C = 6\pi$$

**Step 3** Multiply by $\pi$.

$$C \approx 19 \text{ m}$$

The circumference is about 19 m.

### Example 2

Find the circumference of a circle with a diameter of 24.0 cm.

**Step 1** You know the formula for the circumference is the diameter times $\pi$.

$$C = \pi(24.0)$$

**Step 2** Multiply the diameter by $\pi$.

$$C \approx 75.4 \text{ cm}$$

The circumference is about 75.4 cm.

**Practice Problem** Find the circumference of a circle with a radius of 19 cm.

**Area of a Circle** The formula for the area of a circle is: $A = \pi r^2$

### Example 1

Find the area of a circle with a radius of 4.0 cm.

**Step 1** $A = \pi(4.0)^2$

**Step 2** Find the square of the radius.

$$A = 16\pi$$

**Step 3** Multiply the square of the radius by $\pi$.

$$A \approx 50 \text{ cm}^2$$

The area of the circle is about 50 cm².

### Example 2

Find the area of a circle with a radius of 225 m.

**Step 1** $A = \pi(225)^2$

**Step 2** Find the square of the radius.

$$A = 50625\pi$$

**Step 3** Multiply the square of the radius by $\pi$.

$$A \approx 159043.1$$

The area of the circle is about 159043.1 m².

### Example 3

Find the area of a circle whose diameter is 20.0 mm.

**Step 1** Remember that the radius is half of the diameter.

$$A = \pi\left(\frac{20.0}{2}\right)^2$$

**Step 2** Find the radius.

$$A = \pi(10.0)^2$$

**Step 3** Find the square of the radius.

$$A = 100\pi$$

**Step 4** Multiply the square of the radius by $\pi$.

$$A \approx 314 \text{ mm}^2$$

The area of the circle is about 314 mm².

**Practice Problem** Find the area of a circle with a radius of 16 m.

**Volume** The measure of space occupied by a solid is the **volume** (*V*). To find the volume of a rectangular solid multiply the length times width times height, or $V = l \times w \times h$. It is measured in cubic units, such as cubic centimeters ($cm^3$).

### Example

Find the volume of a rectangular solid with a length of 2.0 m, a width of 4.0 m, and a height of 3.0 m.

**Step 1** You know the formula for volume is the length times the width times the height.

$$V = 2.0 \text{ m} \times 4.0 \text{ m} \times 3.0 \text{ m}$$

**Step 2** Multiply the length times the width times the height.

$$V = 24 \text{ m}^3$$

The volume is 24 $m^3$.

**Practice Problem** Find the volume of a rectangular solid that is 8 m long, 4 m wide, and 4 m high.

To find the volume of other solids, multiply the area of the base times the height.

### Example 1

Find the volume of a solid that has a triangular base with a length of 8.0 m and a height of 7.0 m. The height of the entire solid is 15.0 m.

**Step 1** You know that the base is a triangle, and the area of a triangle is $\frac{1}{2}$ the base times the height, and the volume is the area of the base times the height.

$$V = \left[\frac{1}{2}(b \times h)\right] \times 15$$

**Step 2** Find the area of the base.

$$V = \left[\frac{1}{2}(8 \times 7)\right] \times 15$$

$$V = \left(\frac{1}{2} \times 56\right) \times 15$$

**Step 3** Multiply the area of the base by the height of the solid.

$$V = 28 \times 15$$

$$V = 420 \text{ m}^3$$

The volume is 420 $m^3$.

### Example 2

Find the volume of a cylinder that has a base with a radius of 12.0 cm, and a height of 21.0 cm.

**Step 1** You know that the base is a circle, and the area of a circle is the square of the radius times π, and the volume is the area of the base times the height.

$$V = (\pi r^2) \times 21$$

$$V = (\pi 12^2) \times 21$$

**Step 2** Find the area of the base.

$$V = 144\pi \times 21$$

$$V = 452 \times 21$$

**Step 3** Multiply the area of the base by the height of the solid.

$$V \approx 9{,}500 \text{ cm}^3$$

The volume is about 9,500 $cm^3$.

### Example 3

Find the volume of a cylinder that has a diameter of 15 mm and a height of 4.8 mm.

**Step 1** You know that the base is a circle with an area equal to the square of the radius times π. The radius is one-half the diameter. The volume is the area of the base times the height.

$$V = (\pi r^2) \times 4.8$$

$$V = \left[\pi\left(\frac{1}{2} \times 15\right)^2\right] \times 4.8$$

$$V = (\pi 7.5^2) \times 4.8$$

**Step 2** Find the area of the base.

$$V = 56.25\pi \times 4.8$$

$$V \approx 176.71 \times 4.8$$

**Step 3** Multiply the area of the base by the height of the solid.

$$V \approx 848.2$$

The volume is about 848.2 $mm^3$.

**Practice Problem** Find the volume of a cylinder with a diameter of 7 cm in the base and a height of 16 cm.

# Science Applications

## Measure in SI

The metric system of measurement was developed in 1795. A modern form of the metric system, called the International System (SI), was adopted in 1960 and provides the standard measurements that all scientists around the world can understand.

The SI system is convenient because unit sizes vary by powers of 10. Prefixes are used to name units. Look at **Table 2** for some common SI prefixes and their meanings.

| Table 2 Common SI Prefixes | | | |
|---|---|---|---|
| **Prefix** | **Symbol** | **Meaning** | |
| *kilo-* | k | 1,000 | thousandth |
| *hecto-* | h | 100 | hundred |
| *deka-* | da | 10 | ten |
| *deci-* | d | 0.1 | tenth |
| *centi-* | c | 0.01 | hundreth |
| *milli-* | m | 0.001 | thousandth |

### Example

How many grams equal one kilogram?

**Step 1** Find the prefix *kilo-* in **Table 2.**

**Step 2** Using **Table 2,** determine the meaning of *kilo-*. According to the table, it means 1,000. When the prefix *kilo-* is added to a unit, it means that there are 1,000 of the units in a "kilounit."

**Step 3** Apply the prefix to the units in the question. The units in the question are grams. There are 1,000 grams in a kilogram.

**Practice Problem** Is a milligram larger or smaller than a gram? How many of the smaller units equal one larger unit? What fraction of the larger unit does one smaller unit represent?

## Dimensional Analysis

**Convert SI Units** In science, quantities such as length, mass, and time sometimes are measured using different units. A process called dimensional analysis can be used to change one unit of measure to another. This process involves multiplying your starting quantity and units by one or more conversion factors. A conversion factor is a ratio equal to one and can be made from any two equal quantities with different units. If 1,000 mL equal 1 L then two ratios can be made.

$$\frac{1,000 \text{ mL}}{1 \text{ L}} = \frac{1 \text{ L}}{1,000 \text{ mL}} = 1$$

One can convert between units in the SI system by using the equivalents in **Table 2** to make conversion factors.

### Example

How many cm are in 4 m?

**Step 1** Write conversion factors for the units given. From **Table 2,** you know that 100 cm = 1 m. The conversion factors are

$$\frac{100 \text{ cm}}{1 \text{ m}} \text{ and } \frac{1 \text{ m}}{100 \text{ cm}}$$

**Step 2** Decide which conversion factor to use. Select the factor that has the units you are converting from (m) in the denominator and the units you are converting to (cm) in the numerator.

$$\frac{100 \text{ cm}}{1 \text{ m}}$$

**Step 3** Multiply the starting quantity and units by the conversion factor. Cancel the starting units with the units in the denominator. There are 400 cm in 4 m.

$$4 \text{ m} = \frac{100 \text{ cm}}{1 \text{ m}} = 400 \text{ cm}$$

**Practice Problem** How many milligrams are in one kilogram? (Hint: You will need to use two conversion factors from **Table 2.**)

## Table 3  Unit System Equivalents

| Type of Measurement | Equivalent |
|---|---|
| Length | 1 in = 2.54 cm<br>1 yd = 0.91 m<br>1 mi = 1.61 km |
| Mass and weight* | 1 oz = 28.35 g<br>1 lb = 0.45 kg<br>1 ton (short) = 0.91 tonnes (metric tons)<br>1 lb = 4.45 N |
| Volume | $1\ in^3 = 16.39\ cm^3$<br>1 qt = 0.95 L<br>1 gal = 3.78 L |
| Area | $1\ in^2 = 6.45\ cm^2$<br>$1\ yd^2 = 0.83\ m^2$<br>$1\ mi^2 = 2.59\ km^2$<br>1 acre = 0.40 hectares |
| Temperature | $°C = \dfrac{(°F - 32)}{1.8}$<br>$K = °C + 273$ |

*Weight is measured in standard Earth gravity.

**Convert Between Unit Systems**  **Table 3** gives a list of equivalents that can be used to convert between English and SI units.

### Example

If a meterstick has a length of 100 cm, how long is the meterstick in inches?

**Step 1**  Write the conversion factors for the units given. From **Table 3,** 1 in = 2.54 cm.

$$\frac{1\ in}{2.54\ cm}\ and\ \frac{2.54\ cm}{1\ in}$$

**Step 2**  Determine which conversion factor to use. You are converting from cm to in. Use the conversion factor with cm on the bottom.

$$\frac{1\ in}{2.54\ cm}$$

**Step 3**  Multiply the starting quantity and units by the conversion factor. Cancel the starting units with the units in the denominator. Round your answer to the nearest tenth.

$$100\ \cancel{cm} \times \frac{1\ in}{2.54\ \cancel{cm}} = 39.37\ in$$

The meterstick is about 39.4 in long.

**Practice Problem 1**  A book has a mass of 5 lb. What is the mass of the book in kg?

**Practice Problem 2**  Use the equivalent for in and cm (1 in = 2.54 cm) to show how $1\ in^3 \approx 16.39\ cm^3$.

## Precision and Significant Digits

When you make a measurement, the value you record depends on the precision of the measuring instrument. This precision is represented by the number of significant digits recorded in the measurement. When counting the number of significant digits, all digits are counted except zeros at the end of a number with no decimal point such as 2,050, and zeros at the beginning of a decimal such as 0.03020. When adding or subtracting numbers with different precision, round the answer to the smallest number of decimal places of any number in the sum or difference. When multiplying or dividing, the answer is rounded to the smallest number of significant digits of any number being multiplied or divided.

### Example

The lengths 5.28 and 5.2 are measured in meters. Find the sum of these lengths and record your answer using the correct number of significant digits.

**Step 1**  Find the sum.

| 5.28 m | 2 digits after the decimal |
| + 5.2 m | 1 digit after the decimal |
| 10.48 m | |

**Step 2**  Round to one digit after the decimal because the least number of digits after the decimal of the numbers being added is 1.

The sum is 10.5 m.

**Practice Problem 1**  How many significant digits are in the measurement 7,071,301 m? How many significant digits are in the measurement 0.003010 g?

**Practice Problem 2**  Multiply 5.28 and 5.2 using the rule for multiplying and dividing. Record the answer using the correct number of significant digits.

## Scientific Notation

Many times numbers used in science are very small or very large. Because these numbers are difficult to work with scientists use scientific notation. To write numbers in scientific notation, move the decimal point until only one non-zero digit remains on the left. Then count the number of places you moved the decimal point and use that number as a power of ten. For example, the average distance from the Sun to Mars is 227,800,000,000 m. In scientific notation, this distance is $2.278 \times 10^{11}$ m. Because you moved the decimal point to the left, the number is a positive power of ten.

The mass of an electron is about 0.000 000 000 000 000 000 000 000 000 000 911 kg. Expressed in scientific notation, this mass is $9.11 \times 10^{-31}$ kg. Because the decimal point was moved to the right, the number is a negative power of ten.

### Example

Earth is 149,600,000 km from the Sun. Express this in scientific notation.

**Step 1**  Move the decimal point until one non-zero digit remains on the left.

1.496 000 00

**Step 2**  Count the number of decimal places you have moved. In this case, eight.

**Step 2**  Show that number as a power of ten, $10^8$.

Earth is $1.496 \times 10^8$ km from the Sun.

**Practice Problem 1**  How many significant digits are in 149,600,000 km? How many significant digits are in $1.496 \times 10^8$ km?

**Practice Problem 2**  Parts used in a high performance car must be measured to $7 \times 10^{-6}$ m. Express this number as a decimal.

**Practice Problem 3**  A CD is spinning at 539 revolutions per minute. Express this number in scientific notation.

# Make and Use Graphs

Data in tables can be displayed in a graph—a visual representation of data. Common graph types include line graphs, bar graphs, and circle graphs.

**Line Graph** A line graph shows a relationship between two variables that change continuously. The independent variable is changed and is plotted on the x-axis. The dependent variable is observed, and is plotted on the y-axis.

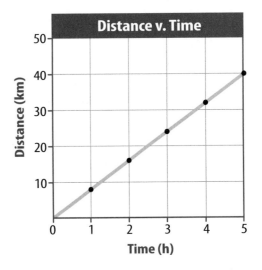

**Figure 8** This line graph shows the relationship between distance and time during a bicycle ride.

### Example

Draw a line graph of the data below from a cyclist in a long-distance race.

| Table 4 Bicycle Race Data | |
|---|---|
| **Time (h)** | **Distance (km)** |
| 0 | 0 |
| 1 | 8 |
| 2 | 16 |
| 3 | 24 |
| 4 | 32 |
| 5 | 40 |

**Step 1** Determine the x-axis and y-axis variables. Time varies independently of distance and is plotted on the x-axis. Distance is dependent on time and is plotted on the y-axis.

**Step 2** Determine the scale of each axis. The x-axis data ranges from 0 to 5. The y-axis data ranges from 0 to 50.

**Step 3** Using graph paper, draw and label the axes. Include units in the labels.

**Step 4** Draw a point at the intersection of the time value on the x-axis and corresponding distance value on the y-axis. Connect the points and label the graph with a title, as shown in **Figure 8.**

**Practice Problem** A puppy's shoulder height is measured during the first year of her life. The following measurements were collected: (3 mo, 52 cm), (6 mo, 72 cm), (9 mo, 83 cm), (12 mo, 86 cm). Graph this data.

**Find a Slope** The slope of a straight line is the ratio of the vertical change, rise, to the horizontal change, run.

$$\text{Slope} = \frac{\text{vertical change (rise)}}{\text{horizontal change (run)}} = \frac{\text{change in } y}{\text{change in } x}$$

### Example

Find the slope of the graph in **Figure 8**.

**Step 1** You know that the slope is the change in y divided by the change in x.

$$\text{Slope} = \frac{\text{change in } y}{\text{change in } x}$$

**Step 2** Determine the data points you will be using. For a straight line, choose the two sets of points that are the farthest apart.

$$\text{Slope} = \frac{(40 - 0) \text{ km}}{(5 - 0) \text{ h}}$$

**Step 3** Find the change in y and x.

$$\text{Slope} = \frac{40 \text{ km}}{5 \text{ h}}$$

**Step 4** Divide the change in y by the change in x.

$$\text{Slope} = \frac{8 \text{ km}}{\text{h}}$$

The slope of the graph is 8 km/h.

**Bar Graph** To compare data that does not change continuously you might choose a bar graph. A bar graph uses bars to show the relationships between variables. The *x*-axis variable is divided into parts. The parts can be numbers such as years, or a category such as a type of animal. The *y*-axis is a number and increases continuously along the axis.

### Example

A recycling center collects 4.0 kg of aluminum on Monday, 1.0 kg on Wednesday, and 2.0 kg on Friday. Create a bar graph of this data.

**Step 1** Select the *x*-axis and *y*-axis variables. The measured numbers (the masses of aluminum) should be placed on the *y*-axis. The variable divided into parts (collection days) is placed on the *x*-axis.

**Step 2** Create a graph grid like you would for a line graph. Include labels and units.

**Step 3** For each measured number, draw a vertical bar above the *x*-axis value up to the *y*-axis value. For the first data point, draw a vertical bar above Monday up to 4.0 kg.

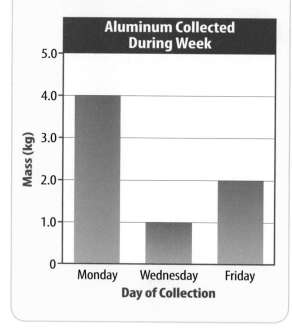

**Practice Problem** Draw a bar graph of the gases in air: 78% nitrogen, 21% oxygen, 1% other gases.

**Circle Graph** To display data as parts of a whole, you might use a circle graph. A circle graph is a circle divided into sections that represent the relative size of each piece of data. The entire circle represents 100%, half represents 50%, and so on.

### Example

Air is made up of 78% nitrogen, 21% oxygen, and 1% other gases. Display the composition of air in a circle graph.

**Step 1** Multiply each percent by 360° and divide by 100 to find the angle of each section in the circle.

$$78\% \times \frac{360°}{100} = 280.8°$$

$$21\% \times \frac{360°}{100} = 75.6°$$

$$1\% \times \frac{360°}{100} = 3.6°$$

**Step 2** Use a compass to draw a circle and to mark the center of the circle. Draw a straight line from the center to the edge of the circle.

**Step 3** Use a protractor and the angles you calculated to divide the circle into parts. Place the center of the protractor over the center of the circle and line the base of the protractor over the straight line.

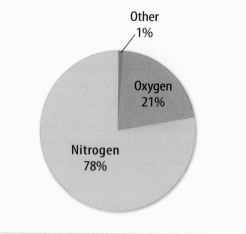

**Practice Problem** Draw a circle graph to represent the amount of aluminum collected during the week shown in the bar graph to the left.

## PERIODIC TABLE OF THE ELEMENTS

Element — Hydrogen
Atomic number — 1
Symbol — **H**
Atomic mass — 1.01
— State of matter

- Gas
- Liquid
- Solid
- Synthetic

A column in the periodic table is called a **group**.

The number in parentheses is the mass number of the longest lived isotope for that element.

A row in the periodic table is called a **period**.

| | 1 | 2 | 3 | 4 | 5 | 6 | 7 | 8 | 9 |
|---|---|---|---|---|---|---|---|---|---|
| **1** | Hydrogen 1 **H** 1.01 | | | | | | | | |
| **2** | Lithium 3 **Li** 6.94 | Beryllium 4 **Be** 9.01 | | | | | | | |
| **3** | Sodium 11 **Na** 22.99 | Magnesium 12 **Mg** 24.31 | | | | | | | |
| **4** | Potassium 19 **K** 39.10 | Calcium 20 **Ca** 40.08 | Scandium 21 **Sc** 44.96 | Titanium 22 **Ti** 47.87 | Vanadium 23 **V** 50.94 | Chromium 24 **Cr** 52.00 | Manganese 25 **Mn** 54.94 | Iron 26 **Fe** 55.85 | Cobalt 27 **Co** 58.93 |
| **5** | Rubidium 37 **Rb** 85.47 | Strontium 38 **Sr** 87.62 | Yttrium 39 **Y** 88.91 | Zirconium 40 **Zr** 91.22 | Niobium 41 **Nb** 92.91 | Molybdenum 42 **Mo** 95.96 | Technetium 43 **Tc** (98) | Ruthenium 44 **Ru** 101.07 | Rhodium 45 **Rh** 102.91 |
| **6** | Cesium 55 **Cs** 132.91 | Barium 56 **Ba** 137.33 | Lanthanum 57 **La** 138.91 | Hafnium 72 **Hf** 178.49 | Tantalum 73 **Ta** 180.95 | Tungsten 74 **W** 183.84 | Rhenium 75 **Re** 186.21 | Osmium 76 **Os** 190.23 | Iridium 77 **Ir** 192.22 |
| **7** | Francium 87 **Fr** (223) | Radium 88 **Ra** (226) | Actinium 89 **Ac** (227) | Rutherfordium 104 **Rf** (267) | Dubnium 105 **Db** (268) | Seaborgium 106 **Sg** (271) | Bohrium 107 **Bh** (272) | Hassium 108 **Hs** (270) | Meitnerium 109 **Mt** (276) |

**Lanthanide series**

| Cerium 58 **Ce** 140.12 | Praseodymium 59 **Pr** 140.91 | Neodymium 60 **Nd** 144.24 | Promethium 61 **Pm** (145) | Samarium 62 **Sm** 150.36 | Europium 63 **Eu** 151.96 |
|---|---|---|---|---|---|

**Actinide series**

| Thorium 90 **Th** 232.04 | Protactinium 91 **Pa** 231.04 | Uranium 92 **U** 238.03 | Neptunium 93 **Np** (237) | Plutonium 94 **Pu** (244) | Americium 95 **Am** (243) |
|---|---|---|---|---|---|

SCIENCE SKILL HANDBOOK

MATH SKILL HANDBOOK

REFERENCE HANDBOOK

GLOSSARY/ GLOSARIO

INDEX

Metal

Metalloid

Nonmetal

Recently discovered

**18**

Helium
2 🎈
**He**
4.00

**13**

**14**

**15**

**16**

**17**

Boron
5 ⬜
**B**
10.81

Carbon
6 ⬜
**C**
12.01

Nitrogen
7 🎈
**N**
14.01

Oxygen
8 🎈
**O**
16.00

Fluorine
9 🎈
**F**
19.00

Neon
10 🎈
**Ne**
20.18

Aluminum
13 ⬜
**Al**
26.98

Silicon
14 ⬜
**Si**
28.09

Phosphorus
15 ⬜
**P**
30.97

Sulfur
16 ⬜
**S**
32.07

Chlorine
17 🎈
**Cl**
35.45

Argon
18 🎈
**Ar**
39.95

**10**

**11**

**12**

Nickel
28 ⬜
**Ni**
58.69

Copper
29 ⬜
**Cu**
63.55

Zinc
30 ⬜
**Zn**
65.38

Gallium
31 ⬜
**Ga**
69.72

Germanium
32 ⬜
**Ge**
72.64

Arsenic
33 ⬜
**As**
74.92

Selenium
34 ⬜
**Se**
78.96

Bromine
35 💧
**Br**
79.90

Krypton
36 🎈
**Kr**
83.80

Palladium
46 ⬜
**Pd**
106.42

Silver
47 ⬜
**Ag**
107.87

Cadmium
48 ⬜
**Cd**
112.41

Indium
49 ⬜
**In**
114.82

Tin
50 ⬜
**Sn**
118.71

Antimony
51 ⬜
**Sb**
121.76

Tellurium
52 ⬜
**Te**
127.60

Iodine
53 ⬜
**I**
126.90

Xenon
54 🎈
**Xe**
131.29

Platinum
78 ⬜
**Pt**
195.08

Gold
79 ⬜
**Au**
196.97

Mercury
80 💧
**Hg**
200.59

Thallium
81 ⬜
**Tl**
204.38

Lead
82 ⬜
**Pb**
207.20

Bismuth
83 ⬜
**Bi**
208.98

Polonium
84 ⬜
**Po**
(209)

Astatine
85 ⬜
**At**
(210)

Radon
86 🎈
**Rn**
(222)

Darmstadtium
110 ⊙
**Ds**
(281)

Roentgenium
111 ⊙
**Rg**
(280)

Copernicium
112 ⊙
**Cn**
(285)

Ununtrium
* 113 ⊙
**Uut**
(284)

Ununquadium
* 114 ⊙
**Uuq**
(289)

Ununpentium
* 115 ⊙
**Uup**
(288)

Ununhexium
* 116 ⊙
**Uuh**
(293)

Ununoctium
* 118 ⊙
**Uuo**
(294)

**\*** The names and symbols for elements 113-116 and 118 are temporary. Final names will be selected when the elements' discoveries are verified.

Gadolinium
64 ⬜
**Gd**
157.25

Terbium
65 ⬜
**Tb**
158.93

Dysprosium
66 ⬜
**Dy**
162.50

Holmium
67 ⬜
**Ho**
164.93

Erbium
68 ⬜
**Er**
167.26

Thulium
69 ⬜
**Tm**
168.93

Ytterbium
70 ⬜
**Yb**
173.05

Lutetium
71 ⬜
**Lu**
174.97

Curium
96 ⊙
**Cm**
(247)

Berkelium
97 ⊙
**Bk**
(247)

Californium
98 ⊙
**Cf**
(251)

Einsteinium
99 ⊙
**Es**
(252)

Fermium
100 ⊙
**Fm**
(257)

Mendelevium
101 ⊙
**Md**
(258)

Nobelium
102 ⊙
**No**
(259)

Lawrencium
103 ⊙
**Lr**
(262)

# Glossary/Glosario

A science multilingual glossary is available on the science Web site. The glossary includes the following languages.

| | | |
|---|---|---|
| Arabic | Hmong | Tagalog |
| Bengali | Korean | Urdu |
| Chinese | Portuguese | Vietnamese |
| English | Russian | |
| Haitian Creole | Spanish | |

**Cómo usar el glosario en español:**
1. Busca el término en inglés que desees encontrar.
2. El término en español, junto con la definición, se encuentran en la columna de la derecha.

## Pronunciation Key
Use the following key to help you sound out words in the glossary.

| | | | | |
|---|---|---|---|---|
| **a** . . . . . . . . . . b**a**ck (BAK) | | | **ew** . . . . . . . . . f**oo**d (FEWD) | |
| **ay** . . . . . . . . . d**ay** (DAY) | | | **yoo** . . . . . . . . p**u**re (PYOOR) | |
| **ah** . . . . . . . . . f**a**ther (FAH thur) | | | **yew** . . . . . . . . f**ew** (FYEW) | |
| **ow** . . . . . . . . fl**ow**er (FLOW ur) | | | **uh** . . . . . . . . . comm**a** (CAH muh) | |
| **ar** . . . . . . . . . c**ar** (CAR) | | | **u (+ con)** . . . . r**u**b (RUB) | |
| **e** . . . . . . . . . . l**e**ss (LES) | | | **sh** . . . . . . . . . **sh**elf (SHELF) | |
| **ee** . . . . . . . . . l**ea**f (LEEF) | | | **ch** . . . . . . . . . na**t**ure (NAY chur) | |
| **ih** . . . . . . . . . tr**i**p (TRIHP) | | | **g** . . . . . . . . . . **g**ift (GIHFT) | |
| **i (i + con + e)** **i**dea (i DEE uh) | | | **j** . . . . . . . . . . **g**em (JEM) | |
| **oh** . . . . . . . . . g**o** (GOH) | | | **ing** . . . . . . . . s**ing** (SING) | |
| **aw** . . . . . . . . s**o**ft (SAWFT) | | | **zh** . . . . . . . . . vi**s**ion (VIH zhun) | |
| **or** . . . . . . . . . **or**bit (OR buht) | | | **k** . . . . . . . . . . ca**k**e (KAYK) | |
| **oy** . . . . . . . . . c**oi**n (COYN) | | | **s** . . . . . . . . . . **s**eed, **c**ent (SEED, SENT) | |
| **oo** . . . . . . . . . f**oo**t (FOOT) | | | **z** . . . . . . . . . . **z**one, rai**s**e (ZOHN, RAYZ) | |

| English | **A** | Español |
|---|---|---|

**acceleration/balanced forces**

**acceleration:** a measure of the change in velocity during a period of time. (p. 27)

**Archimedes' (ar kuh MEE deez) principle:** principle that states that the buoyant force on an object is equal to the weight of the fluid that the object displaces. (p. 134)

**atmospheric pressure:** the ratio of the weight of all the air above you to your surface area. (p. 126)

**average speed:** the total distance traveled divided by the total time taken to travel that distance. (p. 19)

**aceleración/fuerzas en equilibrio**

**aceleración:** medida del cambio de velocidad durante un período de tiempo. (pág. 27)

**principio de Arquímedes:** principio que establece que la fuerza de empuje ejercida sobre un objeto es igual al peso del fluido que el objeto desplaza. (pág. 134)

**presión atmosférica:** peso del aire sobre una superficie. (pág. 126)

**rapidez promedio:** distancia total recorrida dividida por el tiempo usado para recorrerla. (pág. 19)

**B**

**balanced forces:** forces acting on an object that combine and form a net force of zero. (p. 56)

**fuerzas en equilibrio:** fuerzas que actúan sobre un objeto, se combinan y forman una fuerza neta de cero. (pág. 56)

**Bernoulli's (ber NEW leez) principle:** principle that states that the pressure of a fluid decreases when the speed of that fluid increases. (p. 142)

**buoyant (BOY unt) force:** an upward force applied by a fluid on an object in the fluid. (p. 132)

**principio de Bernoulli:** principio que establece que la presión de un fluido disminuye cuando la rapidez de dicho fluido aumenta. (pág. 142)

**fuerza de empuje:** fuerza ascendente que un fluido aplica a un objeto que se encuentra en él. (pág. 132)

**C**

**centripetal (sen TRIH puh tuhl) force:** in circular motion, a force that acts perpendicular to the direction of motion, toward the center of the curve. (p. 66)

**circular motion:** any motion in which an object is moving along a curved path. (p. 66)

**constant speed:** the rate of change of position in which the same distance is traveled each second. (p. 18)

**contact force:** a push or a pull on one object by another object that is touching it. (p. 45)

**fuerza centrípeta:** en movimiento circular, la fuerza que actúa de manera perpendicular a la dirección del movimiento, hacia el centro de la curva. (pág. 66)

**movimiento circular:** cualquier movimiento en el cual un objeto se mueve a lo largo de una trayectoria curva. (pág. 66)

**velocidad constante:** velocidad a la que se cambia de posición, en la cual se recorre la misma distancia por segundo. (pág. 18)

**fuerza de contacto:** empuje o arrastre ejercido sobre un objeto por otro que lo está tocando. (pág. 45)

**D**

**displacement:** the difference between the initial, or starting, position and the final position of an object that has moved. (p. 13)

**drag force:** a force that opposes the motion of an object through a fluid. (p. 144)

**desplazamiento:** diferencia entre la posición inicial, o salida, y la final de un objeto que se ha movido. (pág. 13)

**fuerza de arrastre:** fuerza que se opone al movimiento de un objeto a través de un fluido. (pág. 144)

**E**

**efficiency:** the ratio of output work to input work. (p. 99)

**eficiencia:** relación entre energía invertida y energía útil. (pág. 99)

**F**

**fluid:** any substance that can flow and take the shape of the container that holds it. (p. 123)

**force:** a push or a pull on an object. (p. 45)

**force pair:** the forces two objects apply to each other. (p. 71)

**friction:** contact force that resists the sliding motion of two touching surfaces. (p. 49)

**fluido:** cualquier sustancia que puede fluir y toma la forma del recipiente que lo contiene. (pág. 123)

**fuerza:** empuje o arrastre ejercido sobre un objeto. (pág. 45)

**par de fuerzas:** fuerzas que dos objetos se aplican entre sí. (pág. 71)

**fricción:** fuerza que resiste el movimiento de dos superficies que están en contacto. (pág. 49)

SCIENCE SKILL HANDBOOK

MATH SKILL HANDBOOK

REFERENCE HANDBOOK

GLOSSARY/ GLOSARIO

INDEX

**fulcrum:** the point about which a lever pivots. (p. 104)

**fulcro:** punto alrededor del cual gira una palanca. (pág. 104)

**gravity:** an attractive force that exists between all objects that have mass. (p. 47)

**gravedad:** fuerza de atracción que existe entre todos los objetos que tienen masa. (pág. 47)

**inclined plane:** a simple machine that consists of a ramp, or a flat, sloped surface. (p. 107)

**plano inclinado:** máquina simple que consiste en una rampa, o superficie plana inclinada. (pág. 107)

**inertia (ihn UR shuh):** the tendency of an object to resist a change in its motion. (p. 58)

**inercia:** tendencia de un objeto a resistirse al cambio en su movimiento. (pág. 58)

**instantaneous speed:** an object's speed at a specific instant in time. (p. 18)

**velocidad instantánea:** velocidad de un objeto en un instante específico en el tiempo. (pág. 18)

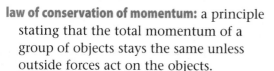

**law of conservation of momentum:** a principle stating that the total momentum of a group of objects stays the same unless outside forces act on the objects. (p. 74)

**ley de la conservación del momentum:** principio que establece que el momentum total de un grupo de objetos permanece constante a menos que fuerzas externas actúen sobre los objetos. (pág. 74)

**lever:** a simple machine that consists of a bar that pivots, or rotates, around a fixed point. (p. 104)

**palanca:** máquina simple que consiste en una barra que gira, o rota, alrededor de un punto fijo. (pág. 104)

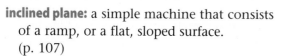

**mass:** the amount of matter in an object. (p. 47)

**masa:** cantidad de materia en un objeto. (pág. 47)

**mechanical advantage:** the ratio of a machine's output force produced to the input force applied. (p. 98)

**ventaja mecánica:** relación entre la fuerza útil que produce una máquina con la fuerza aplicada. (pág. 98)

**momentum:** a measure of how hard it is to stop a moving object. (p. 73)

**momentum:** medida de qué tan difícil es detener un objeto en movimiento. (pág. 73)

**motion:** the process of changing position. (p. 13)

**movimiento:** proceso de cambiar de posición. (pág. 13)

**net force:** the combination of all the forces acting on an object. (p. 55)

**fuerza neta:** combinación de todas las fuerzas que actúan sobre un objeto. (pág. 55)

**Newton's first law of motion:** law that states that if the net force acting on an object is zero, the motion of the object does not change. (p. 57)

**primera ley del movimiento de Newton:** ley que establece que si la fuerza neta ejercida sobre un objeto es cero, el movimiento de dicho objeto no cambia. (pág. 57)

**Newton's second law of motion:** law that states that the acceleration of an object is equal to the net force exerted on the object divided by the object's mass. (p. 65)

**Newton's third law of motion:** law that states that for every action there is an equal and opposite reaction. (p. 71)

**noncontact force:** a force that one object applies to another object without touching it. (p. 46)

**segunda ley del movimiento de Newton:** ley que establece que la aceleración de un objeto es igual a la fuerza neta que actúa sobre él divida por su masa. (pág. 65)

**tercera ley del movimiento de Newton:** ley que establece que para cada acción hay una reacción igual en dirección opuesta. (pág. 71)

**fuerza de no contacto:** fuerza que un objeto puede aplicar sobre otro sin tocarlo. (pág. 46)

**P**

**Pascal's (pas KALZ) principle:** principle that states that when pressure is applied to a fluid in a closed container, the pressure increases by the same amount everywhere in the container. (p. 140)

**position:** an object's distance and direction from a reference point. (p. 9)

**power:** the rate at which work is done. (p. 91)

**pressure:** the amount of force per unit area applied to an object's surface. (p. 124)

**pulley:** a simple machine that consists of a grooved wheel with a rope or cable wrapped around it. (p. 109)

**principio de Pascal:** principio que establece que cuando se aplica presión a un fluido en un recipiente cerrado, la presión aumenta el mismo valor en todas las partes del recipiente. (pág. 140)

**posición:** distancia y dirección de un objeto según un punto de referencia. (pág. 9)

**potencia:** velocidad a la que se hace trabajo. (pág. 91)

**presión:** cantidad de fuerza por unidad de área aplicada a la superficie de un objeto. (pág. 124)

**polea:** máquina simple que consiste en una rueda acanalada rodeada por una cuerda o cable. (pág. 109)

**R**

**reference point:** the starting point you use to describe the motion or the position of an object. (p. 9)

**punto de referencia:** punto que se escoge para describir la ubicación, o posición, de un objeto. (pág. 9)

**S**

**screw:** a simple machine that consists of an inclined plane wrapped around a cylinder. (p. 108)

**simple machine:** a machine that does work using one movement. (p. 103)

**speed:** the distance an object moves divided by the time it takes to move that distance. (p. 17)

**tornillo:** máquina simple que consiste en un plano inclinado incrustado alrededor de un cilindro. (pág. 108)

**máquina simple:** máquina que hace trabajo con un movimiento. (pág. 103)

**rapidez:** distancia que un objeto recorre dividida por el tiempo que éste tarda en recorrer dicha distancia. (pág. 17)

**U**

**unbalanced forces:** forces acting on an object that combine and form a net force that is not zero. (p. 56)

**fuerzas no balanceadas:** fuerzas que actúan sobre un objeto, se combinan y forman una fuerza neta diferente de cero. (pág. 56)

SCIENCE SKILL HANDBOOK

MATH SKILL HANDBOOK

REFERENCE HANDBOOK

GLOSSARY/ GLOSARIO

INDEX

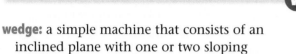

**velocity:** the speed and the direction of a moving object. (p. 23)

**velocidad:** rapidez y dirección de un objeto en movimiento. (pág. 23)

**wedge:** a simple machine that consists of an inclined plane with one or two sloping sides; it is used to split or separate an object. (p. 108)

**weight:** the gravitational force exerted on an object. (p. 48)

**wheel and axle:** a simple machine that consists of an axle attached to the center of a larger wheel, so that the shaft and wheel rotate together. (p. 106)

**work:** the amount of energy used as a force moves an object over a distance. (p. 87)

**cuña:** máquina simple que consiste en un plano inclinado con uno o dos lados inclinados; se usa para partir o separar un objeto. (pág. 108)

**peso:** fuerza gravitacional ejercida sobre un objeto. (pág. 48)

**rueda y eje:** máquina simple que consiste en un eje insertado en el centro de una rueda grande, de manera que el eje y la rueda rotan juntos. (pág. 106)

**trabajo:** cantidad de energía usada como fuerza que mueve un objeto a cierta distancia. (pág. 87)

SCIENCE SKILL HANDBOOK

MATH SKILL HANDBOOK

REFERENCE HANDBOOK

GLOSSARY/ GLOSARIO

INDEX

# Index

SCIENCE SKILL HANDBOOK

MATH SKILL HANDBOOK

REFERENCE HANDBOOK

GLOSSARY/ GLOSARIO

INDEX

# Credits

## Photo Credits

**Cover** Matt Meadows/Peter Arnold, Inc; **ConnectED** (t)Richard Hutchings, (c)Getty Images, (b)Jupiter Images/Thinkstock/Alamy; **ix** (b)Fancy Photography/Veer; **4** (t)Wildscape/Alamy, (b)Andy Crawford/BBC Visual Effects—modelmaker/Dorling Kindersley; **5** (t)Miriam Maslo/Science Photo Library/Corbis, (b)The Gravity Group, LLC; **6–7** Andrew Holt/Getty Images; **7** Camille Moirenc/Getty Images; **8** Hutchings Photography/Digital Light Source; **12** Aerial Photos of New Jersey; **14** Camille Moirenc/Getty Images; **15** David J. Green—technology/Alamy; **16** Heinrich van den Berg/Getty Images; **17** (t to b)Hutchings Photography/Digital Light Source, (2)Department of Defense, (3)Car Culture/Corbis, (4)Andersen Ross/Jupiter Images; **20 Hutchings** Photography/Digital Light Source; **22** Michael Dunning/Photo Researchers; **23–24** Hutchings Photography/Digital Light Source; **25** (r)(bl)(t to b)(2)(3)(4)Richard Hutchings/Digital Light Source, (5)Aaron Haupt; **26** Leo Dennis Productions/Brand X/Corbis; **30** (t)Hutchings Photography/Digital Light Source, (b)Robert Holmes/Corbis; **31** (t)Comstock/PunchStock, (b)Eyecon Images/Alamy Images; **32** (t)NASA, (b)Agence Zoom/Getty Images; **33** Leo Dennis Productions/Brand X/Corbis; **34** Hutchings Photography/Digital Light Source; **36** (t)Camille Moirenc/Getty Images, (c)Steve Bloom Images/Alamy, (b)Agence Zoom/Getty Images; **39** Andrew Holt/Getty Images; **42–43** Jason Horowitz/zefa/Corbis; **44** NASA; **45** (t)Hutchings Photography/Digital Light Source, (b)Terje Rakke/Getty Images; **46** (t)Hutchings Photography/Digital Light Source, (c)Daniel Smith/Getty Images, (b)Ryuhei Shindo/Corbis/Jupiter Images; **48** (l)(r)NASA, (tl)(tr)Horizons Companies; **49** (b)Hutchings Photography/Digital Light Source; **50** Hutchings Photography/Digital Light Source; **51** (t)Terje Rakke/Getty Images, (b)Hutchings Photography/Digital Light Source; **52** (t)NASA/Goddard Space Flight Center Scientific Visualization Studio, (c)VisionsofAmerica/Joe Sohm/Getty Images, (b)Ursula Gahwiler/photolibrary.com, (bkgd)StockTrek/Getty Images; **53** Bryan Mullennix/Getty Images; **54** (t)Hutchings Photography/Digital Light Source, (b)Marvin E. Newman/Getty Images; **55–56** Hutchings Photography/Digital Light Source; **57** (t)Tim Keatley/Alamy, (b)Hutchings Photography/Digital Light Source; **58** AP Photo/Insurance Institute for Highway Safety; **59** (t)Hutchings Photography/Digital Light Source, (c)Tim Keatley/Alamy, (b)AP Photo/Insurance Institute for Highway Safety; **60** (t)Macmillan/McGraw-Hill, (b)Hutchings Photography/Digital Light Source; **61** Michael Steele/Getty Images; **62** Hutchings Photography/Digital Light Source; **63** (l)Tim Garcha/Corbis, (r)Myrleen Ferguson Cate/PhotoEdit; **64** Hutchings Photography/Digital Light Source; **67** Tim Garcha/Corbis; **68** Hutchings Photography/Digital Light Source; **69** Patrik Giardino/Corbis; **70** Hutchings Photography/Digital Light Source; **71** (t)LIN HUI/Xinhua/Landov, (tl)David Madison/Getty Images, (b)Duncan Soar/Alamy; **72** (r)Design Pics Inc./Alamy, (bl)U.S. Air Force photo by Carleton Bailie; **74** (tl)Richard Megna, Fundamental Photographs, NYC, (r)Hutchings Photography/Digital Light Source, (bl)Richard Megna, Fundamental Photographs, NYC; **75** (t)Hutchings Photography/Digital Light Source, (c)David Madison/Getty Images, (b)Richard Megna, Fundamental Photographs, NYC; **76** (tr)Macmillan/McGraw-Hill, (cr)Hutchings Photography/Digital Light Source, (bl)(br)Macmillan/McGraw-Hill, (inset)C Squared Studios/Getty Images, (t to b)Hutchings Photography/Digital Light Source, (2)Macmillan/McGraw-Hill, (3)Hutchings Photography/Digital Light Source, (4)Macmillan/McGraw-Hill; **77** Hutchings Photography/Digital Light Source; **78** (t to b)Hutchings Photography/Digital Light Source, (2)Marvin E. Newman/Getty Images, (3)Myrleen Ferguson Cate/PhotoEdit, (4)LIN HUI/Xinhua/Landov; **81** Jason Horowitz/zefa/Corbis; **84–85** Neil Duncan/photolibrary.com; **86** Malcolm Fife/Getty Images; **87** (t)Hutchings Photography/Digital Light Source, (bl) (br)Jupiterimages/Brand X/Alamy; **88–89** (l)Hutchings Photography/Digital Light Source; **90** Hutchings Photography/Digital Light Source; **92** (tl)Jupiterimages/Brand X/Alamy, (cl)Hutchings Photography/Digital Light Source, (r)Jupiterimages/Brand X/Alamy, (bl)Hutchings Photography/Digital Light Source; **93** (t)P.H. Emerson/George Eastman House/Getty Images, (c)Hulton Archive/Getty Images, (b)Bridgeman Art Library/SuperStock, (inset)Eduardo M. Rivero/age fotostock; **94** Philip and Karen Smith/Getty Images; **95** (t)Hutchings Photography/Digital Light Source, (c)Clive Streeter/Getty Images, (b)Corbis; **96** (t)Hutchings Photography/Digital Light Source, (bl)(br)The McGraw-Hill Companies; **97** (t)Steve Gorton/Dorling Kindersley/Getty Images (c)Sean Justice/Getty Images, (b)Dorling Kindersley; **99** Glowimages/Getty Images; **100** (tl)Clive Streeter/Getty Images, (cl)(cr)The McGraw-Hill Companies, (bl)Glowimages/Getty Images, (br)Sean Justice/Getty Images; **101** (t to b)(2)(3)Hutchings Photography/Digital Light Source, (4)Macmillan/McGraw-Hill; **102** SoloStock Travel/Alamy; **103** (t)McGraw-Hill, (cl)Bob Elsdale, (c)Rod McLean/Alamy, (cr)David Papazian Photography Inc./Jupiterimages, (bl)The McGraw-Hill Companies, (bc)F. Schussler/PhotoLink/Getty Images, (br)Susan E. Degginger/Alamy; **104** (t)Jupiterimages, (c)Digital Vision/Alamy, (b)Andy Aitchison/Corbis; **106** (tl)Matt Carr/Getty Images, (tc)Maurilio Cheli/epa/Corbis, (tr)liquidlibrary/PictureQuest, (b)Jupiterimages/Brand X/Alamy; **108** (t)Mark Douet/Getty Images, (b)Reimar/Alamy; **109** Hutchings Photography/Digital Light Source; **110** (t)imagebroker/Alamy, (b)Brand Z/Alamy; **111** (tl)F. Schussler/PhotoLink/Getty Images, (cl)Digital Vision/Alamy, (r)Corbis, (bl)Mark Douet/Getty Images; **112** (br)(t to b)Hutchings Photography/Digital Light Source, (2)The McGraw-Hill Companies, (3)Macmillan/McGraw-Hill, (4)The McGraw-Hill Companies, (5)Hutchings Photography/Digital Light Source; **113** Hutchings Photography/Digital Light Source; **114** (t)Hutchings Photography/Digital Light Source, (c)Steve Gorton/Dorling Kindersley/Getty Images, (b)Jupiterimages; **117** (l)Hutchings Photography/Digital Light Source, (r)Neil Duncan/photolibrary.com; **SR-0–SR-1** (bkgd)Gallo Images—Neil Overy/Getty Images; **SR-2** Hutchings Photography/Digital Light Source; **SR-6** Michell D. Bridwell/PhotoEdit; **SR-7** (t)The McGraw-Hill Companies, (b)Dominic Oldershaw; **SR-8** StudiOhio; **SR-9** Timothy Fuller; **SR-10** Aaron Haupt; **SR-12** KS Studios; **SR-13 SR-47** Matt Meadows; **SR-48** (c)NIBSC/Photo Researchers, Inc., (r)Science VU/Drs. D.T. John & T.B. Cole/Visuals Unlimited, Inc., Stephen Durr; **SR-49** (t)Mark Steinmetz, (r)Andrew Syred/Science Photo Library/Photo Researchers, (br)Rich Brommer; **SR-50** (l)Lynn Keddie/Photolibrary, (tr)G.R. Roberts, David Fleetham/Visuals Unlimited/Getty Images; **SR-51** Gallo Images/Corbis.